An Introduction to the Geometry of *N* Dimensions

D. M. Y. SOMMERVILLE

Dover Publications, Inc.
Mineola, New York

Bibliographical Note

This Dover edition, first published in 2020, is an unabridged republication of the work originally published in 1929 by Methuen & Company, London.

Library of Congress Cataloging-in-Publication Data

Names: Sommerville, Duncan M'Laren Young, 1879-1934, author.
Title: An introduction to the geometry of N dimensions / D. M. Y. Sommerville.
Description: Dover edition [2020 edition]. | Mineola : Dover Publications, Inc., 2020. | An unabridged republication of the work originally published: London : Methuen & Company, London, 1929; republished by Dover Publications in 1958. | Summary: "For many years, this was the only English-language book devoted to the subject of higher-dimensional geometry. While that is no longer the case, it remains a significant contribution to the literature, exploring topics of perennial interest to geometers: the fundamental ideas of incidence, parallelism, perpendicularity, angles between linear spaces, and polytopes. Analytical geometry is examined from the projective and analytic points of view. Includes 60 diagrams"— Provided by publisher.
Identifiers: LCCN 2019038772 | ISBN 9780486842486 (trade paperback)
Subjects: LCSH: Hyperspace. | Geometry.
Classification: LCC QA691 .S6 2020 | DDC 516/.158—dc23
LC record available at https://lccn.loc.gov/2019038772

Manufactured in the United States by LSC Communications
84248701
www.doverpublications.com

2 4 6 8 10 9 7 5 3 1

2019

TO

MY WIFE

PREFACE

IT is scarcely necessary to apologise for writing a book on n-dimensional geometry. One should regret rather the comparative neglect which the subject has suffered at the hands of British mathematicians.* Yet one may almost say that this country was its home of origin, for, with the exception of a few previous sporadic references, the first paper dealing explicitly with geometry of n dimensions was one by Cayley in 1843, and the importance of the subject was recognised from the first by three of our most famous pure mathematicians—Cayley, Clifford, and Sylvester. On the Continent the classical works of Grassmann and Schläfli attracted at first no attention. Schläfli's remarkable memoir, in fact, failed to secure publication, and in spite of Cayley's gallant attempt at rescue by translating and publishing part of it in the "Quarterly Journal" it remained unknown until it was found and published several years after the author's death, and fifty years after it was written. By that time Schlegel and others in Germany had made the subject well known, but mostly in its metrical aspect. The wonderful projective geometry of hyperspace has been almost entirely the product of the gifted Italian school of

* In the twenty-seven volumes of the new series of the Proceedings of the London Mathematical Society there are barely a dozen papers dealing with higher space. On the other hand, it is interesting to notice that there are about an equal number in the three volumes of the Journal; this seems to indicate a revival of interest.

geometers ; though this branch also was inaugurated by a British mathematician, W. K. Clifford, in 1878.

The present introduction deals with the metrical and to a slighter extent with the projective aspect. A third aspect, which has attracted much attention recently, from its application to relativity, is the differential aspect. This is altogether excluded from the present book.

In writing this book I have not attempted to produce a complete systematic treatise, but have rather selected certain representative topics which not only illustrate the extensions of theorems of three-dimensional geometry, but reveal results which are unexpected and where analogy would be a faithless guide.

The first four chapters explain the fundamental ideas of incidence, parallelism, perpendicularity, and angles between linear spaces ; and in Chapter I there is an excursus into enumerative geometry which may be omitted on a first reading. Chapters V and VI are analytical, the former projective, the latter largely metrical. In the former are given some of the simplest ideas relating to algebraic varieties, and a more detailed account of quadrics, especially with reference to their linear spaces. In the latter there are given, in addition to the ordinary cartesian formulæ, some account and applications of the Plücker-Grassmann co-ordinates of a linear space, and applications to line-geometry. The remaining chapters deal with polytopes, and contain, especially in Chapter IX, some of the elementary ideas in analysis situs. Chapter VIII treats of the content of hyperspatial figures, and the final chapter establishes the regular polytopes.

A number of references have been given at the ends of the chapters. Some of these are the original works in which the various theories were first expounded, others are a selection of more recent works in which a fuller account may be found. Reference may be made to the author's "Bibliography of Non-Euclidean Geometry, including the theory of parallels, the foundations of geometry, and space of *n* dimensions" (London: Harrison, for the University of St. Andrews. 1911), which contains, in addition to a chronological catalogue, a detailed subject index and an index of authors.

I am indebted in particular to Schoute's "Mehrdimensionale Geometrie" (Leipzig, 2 vols., 1902 and 1905), Bertini's "Introduzione alla geometria proiettiva degli iperspazi" (Pisa, 1907), and the various articles of the "Encyklopädie der mathematischen Wissenschaften."

For assistance in correcting the proofs I have to thank Mr. F. F. Miles, M.A., Lecturer in Mathematics at this College.

D. M. Y. S.

VICTORIA UNIVERSITY COLLEGE,
WELLINGTON, N.Z., *May*, 1929.

CONTENTS

CHAPTER I

Fundamental Ideas

CHAPTER II

Parallels

CHAPTER III

PERPENDICULARITY

CHAPTER IV

DISTANCES AND ANGLES BETWEEN FLAT SPACES

CHAPTER V

Analytical Geometry: Projective

CHAPTER VI

Analytical Geometry: Metrical

CHAPTER VII

Polytopes

CHAPTER VIII

Mensuration: Content

CHAPTER IX

EULER'S THEOREM

CHAPTER X

THE REGULAR POLYTOPES

An Introduction to the Geometry of *N* Dimensions

GEOMETRY OF N DIMENSIONS

CHAPTER I

FUNDAMENTAL IDEAS

1. Origins of Geometry. Geometry for the individual begins intuitionally and develops by a co-ordination of the senses of sight and touch. Its history followed a similar course. The crude ideas of shape, bulk, superficial extent, and length became analysed, refined, and made abstract, and led to the conception of geometrical figures. The development started with the solid; surface and line, without solidity, were later abstractions. Witness the inability of most animals and some primitive races of men to recognise a picture. Having no depth, except such as is imitated by the skilfulness of the drawing or shading, it conveys to the undeveloped intelligence only an impression of flat regions of contrasted colouring. When the power of abstraction had proceeded to the extent of conceiving surfaces apart from solids, plane geometry arose. The idea of dimensionality was then formed, when a region of two dimensions was recognised within the three-dimensional universe. This stage had been reached when Greek geometry started. It was many centuries, however, before the human mind began to conceive of an upward extension to the idea of dimensionality, and even now this conception is confined to the comparatively very small class of mathematicians and philosophers.

2. Extension of the Dimensional Idea. There are two main ways in which we may arrive at an idea of higher dimensions: one geometrical, by extending in the upward direction the series of geometrical elements, point, line, surface, solid; the other by invoking algebra and giving extended geometrical interpretations to algebraic relationships. In whatever way we may proceed we are led to the invention of new elements which have to be defined strictly and logically if exact deductions are to be made. A great deal is suggested by analogy, but while analogy is often a useful guide and stimulus, it provides no proofs, and may often lead one astray if not supplemented by logical reasoning. If we follow the geometrical method the only safe course is that which was systematically laid down for the first time by Euclid, that is to lay down a basis of axioms or assumptions. When we leave the field of sensuous perception and can no longer depend upon intuition as a guide, our axioms will no longer be "self-evident truths," but simply statements, assumed without proof, as a basis for future deductions.

3. Definitions and Axioms. In geometry there are objects which have to be defined, and relationships between these objects which have to be deduced either from the definitions or from other simpler relationships. In defining an object we must make reference to some simpler object, hence there must be some objects which have to be left undefined, the *indefinables*. Similarly, in deducing relations from simpler ones we must arrive back at certain statements which cannot be deduced from anything simpler; these are the *axioms* or unproved propositions. The whole science of geometry can thus be made to rest upon a set of definitions and axioms. The actual choice of fundamental definitions and axioms is to a certain extent arbitrary, but there are certain principles which have to be considered in making a choice of axioms. These are:—

(1) *Self-consistency.* The set of axioms must be logically self-consistent. No axiom must be in conflict with deductions from any of the other axioms.

(2) *Non-redundance or Independence.* This condition is not a necessary one, but in a logical scheme it is desirable. Pedagogically the condition is frequently ignored.

(3) *Categoricalness.* This means not only that the set of axioms should be complete and sufficient for the development of the science, but that it should be possible to construct only a unique set of entities for which the axioms are valid. It is doubtful whether any set of axioms can be strictly categorical. If any set of entities is constructed so as to satisfy the axioms,

it is nearly always, if not always, possible to change the ideas and construct another set of entities also satisfying the axioms. Thus with the ordinary ideas of point and straight line in plane geometry the axioms can still be applied when instead of a point we substitute a pair of numbers (x, y), and instead of straight line an equation of the first degree in x and y; corresponding to the incidence of a point with a straight line we have the fact that the values of x and y satisfy the equation. It is desirable, in fact, that the set of axioms should *not* be categorical, for thereby they are given a wider field of validity, and propositions proved for the one set of entities can be transferred at once to another set, perhaps in a different branch of mathematics.

4. The Axioms of Incidence. As indefinables we shall choose first the *point*, *straight line*, and *plane*. With regard to these we shall proceed to make certain statements, the axioms. If these should appear to be very obvious, and as if they might be taken for granted, it will be a good corrective for the reader to replace the words point and straight line, which he must remember are not yet defined, by the names of other objects to which the axioms may be made to apply, such as "committee member" and "committee." Following Hilbert we divide the axioms into groups.

THE AXIOMS

Group I

AXIOMS OF INCIDENCE OR CONNECTION

I. 1. *Any two distinct points uniquely determine a straight line.*

We imagine a collection of individuals who have a craze for organisation and form themselves into committees. The committees are so arranged that every person is on a committee along with each of the others, but no two individuals are to be found together on more than one committee.

I. 2. *If A, B are distinct points there is at least one point not on the straight line AB.*

This is an "existence-postulate."

I. 3. *Any three non-collinear points determine a plane.*

I. 4. *If two distinct points A, B both belong to a plane* α, *every point of the straight line AB belongs to* α.

From I. 1 it follows that two distinct straight lines have either one or no point in common. From I. 4 it follows that a straight line and a plane have either no point or one point in common, or else the straight line lies entirely in the plane; from I. 3 and 4 two distinct planes have either no point, one point, or a whole straight line in common.

I. 5. *If A, B, C are non-collinear points there is at least one point not on the plane ABC.*

I. 2 and 5 are existence-postulates; 2 implies two-dimensional geometry, and 5 three-dimensional.

The next of Hilbert's axioms is that if two planes have one point A in common they have a second point B in common, and therefore by I. 4 they have the whole straight line AB in common. If this is assumed it limits space to three dimensions.

5. Projective Geometry. There is a difficulty in determining all the elements of space by means of the existence postulates and other axioms, for while Axiom I. 1 postulates that any two points determine a line, there is no axiom which secures that any two lines in a plane will determine a point. In fact, in euclidean geometry this is not true since parallel lines have no point in common. For the present therefore we shall confine ourselves to a simpler and more symmetrical type of geometry, *projective geometry*, for which we add the following axiom:

I. 1'. *Any two distinct straight lines in a plane uniquely determine a point.*

6. Construction of Three-dimensional Space. We may proceed now to obtain all the elements, points, lines, etc., of space with the help of the existence postulates and other axioms. Starting with two points A, B we determine the line AB, and on this we shall suppose all the points determined. Taking a third point C, not on AB, a plane ABC is determined. In this plane there are determined: first, the three points A, B, C, then the three lines BC, CA, AB and all their points; then all lines determined by two points one on each of two of these lines, and since every line in the plane meets these three lines all the lines of the plane are thus determined; finally any point of the plane is determined by the intersection of two lines which have already been determined, e.g. if P is any point of the plane, the line PA cuts BC in a point L and PB cuts CA in a point M; there are therefore two points, L on BC and M on CA, such that LA and MB determine P.

We next take a point D not in the plane ABC, and determine planes, lines and points as follows: first, the planes DBC, DCA, DAB and all the lines and points in these planes; then the lines determined by joining D to the points of ABC, and all the points on these lines, and all the lines and planes determined by any two or any three of the points thus determined. If now P is any point we cannot be sure that it is one of the points thus determined unless the line DP meets the plane ABC. For this we could assume as an axiom: "every line meets every plane in a point," which is true for projective geometry of three dimensions; but a weaker axiom, which is true also in euclidean geometry, is sufficient, viz. Hilbert's axiom:

I. 6. *If two planes have a point A in common, they have a second point B in common.*

Then if P is any point, the planes PAD and ABC, which have the point A common, have another point, say Q, common, and the lines AQ and DP which lie in the same plane determine a point R which lies in the line DP and also in the plane ABC. Hence DP meets the plane ABC, and therefore all points are obtained by this process.

If l is any line (Fig. 1), determined by the two points P, Q, the planes PQD and ABD, having D in common, intersect in a line which cuts PQ in N, say, and AB in E; then PQ and CE lie in the same plane and therefore meet in a point Z. Hence every line cuts the plane ABC.

7. If α is any plane (Fig. 2), determined by the three points P, Q, R, and O is a point not in this plane, the lines OP, OQ, OR cut ABC in P′, Q′, R′. QR and Q′R′, being in the same plane, intersect in a point X. Similarly RP and R′P′ intersect in Y, and PQ and P′Q′ in Z. Hence every plane cuts the plane ABC.

The points X, Y, Z all lie in both of the planes PQR and P′Q′R′ and are therefore collinear. This theorem is known as *Desargues' Theorem*, PQR and P′Q′R′ being two triangles in perspective. (It is interesting to note in passing that the corresponding theorem for two triangles in perspective *in the same plane* cannot be proved without the assumption of three dimensions or some special axiom in addition to those which we have assumed.)

We can now prove that any two planes intersect in a line. Let α and α' be any two planes. We have seen that every plane cuts ABC in a straight line. Let α and α' cut ABC in the lines l and l'. These lines intersect in a point P which is therefore common to the two planes, and therefore the two planes, having a point in common, intersect in a straight line.

Next, every line cuts every plane in a point. For if β is any

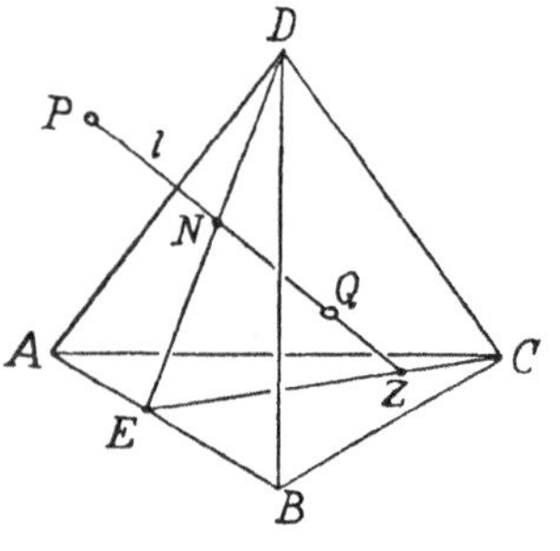

FIG. 1.

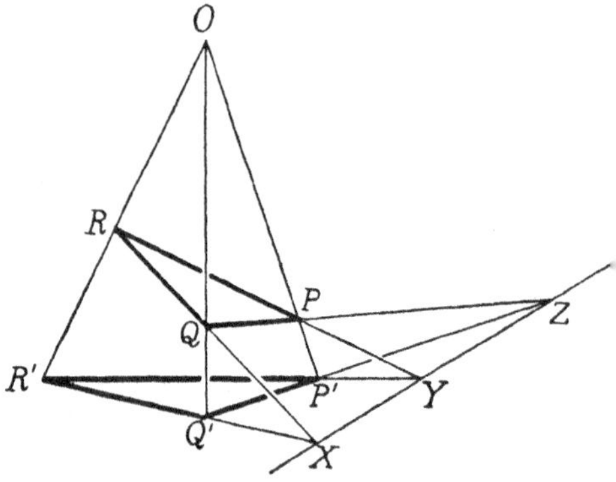

FIG. 2.

plane through the given line l, this plane cuts the given plane α in a line m, and the two lines l and m, lying in the same plane, intersect in a point, which is common to l and α.

Lastly, any three planes have a point in common. For the planes α, β intersect in a line l, and l cuts γ in a point; this point is common to the three planes.

8. Extension to Four Dimensions. All the points, lines and planes determined from four given non-coplanar points form a three-dimensional region, which is ordinary projective space of three dimensions. We proceed to postulate the existence of at least one point E not in this region. The three-dimensional region is not now the whole of space, but will be called a *hyperplane* lying in hyperspace. A hyperplane is thus determined by four points. We may determine similarly the hyperplanes ABCE, ABDE, etc., and the planes, lines and points in them, and further the hyperplanes, planes, and lines determined by four, three, or two points not all lying in one of these hyperplanes. The hyperplanes ABCE and ABDE have the three points A, B, E in common and therefore the plane ABE. We may show that any two hyperplanes in this region have a plane in common. For example, if α and α' are two hyperplanes determined by four points P, Q, R, S and P′, Q′, R′, S′ lying on EA, EB, EC, ED respectively, PQ and P′Q′, being in the plane EAB, cut in a point Z, similarly QR and Q′R′ cut in a point X, and RP and R′P′ in a point Y; PS and P′S′ cut in U, QS and Q′S′ in V, RS and R′S′ in W. The six points X, Y, Z, U, V, W are thus all common to α and α'. These six points form the vertices of a complete quadrilateral (Fig. 3) whose sides are XYZ, the intersection of the planes PQR and P′Q′R′, XVW, the intersection of QRS and Q′R′S′, etc. The six points are thus coplanar, and this plane is common to the two hyperplanes α and α'.

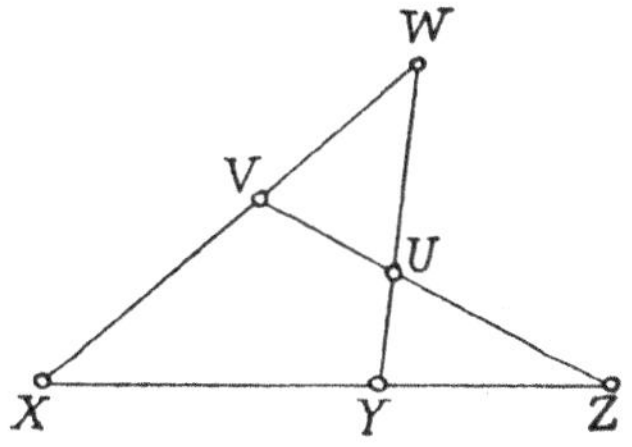

FIG. 3.

Thus in this region, space of four dimensions, two hyperplanes intersect in a plane. Similarly three hyperplanes intersect in a line, four hyperplanes in a point, while five hyperplanes do not in general have any point in common. A hyperplane cuts a plane (intersection of two hyperplanes) in a line, and a line in a point. Two planes, each the intersection of two hyperplanes, have in general only one point in common, a plane and a line in general have no point in common.

A hyperplane is determined by four points, a point and a plane, or by two skew lines.

9. Degrees of Freedom: Dimensions. A point in a line is said to have one degree of freedom, in a plane two, in a hyperplane three, and in the hyperspacial region four. The point being taken as the element, a line is said to be of one dimension, a plane two, a hyperplane three, and the hyperspacial region four. For a point to be in a given hyperplane one condition is required, or one degree of constraint, thus reducing the number of degrees of freedom from four to three. For a point to lie in a plane two conditions are required; if it is to lie simultaneously in two planes four conditions are required and it is completely determined.

10. Extension to n Dimensions. We may now extend these ideas straightaway to n dimensions, and at the same time acquire both greater generality and greater succinctness in expression. The series point, line, plane, hyperplane (or as it is more explicitly termed *three-flat*), . . ., n-flat are regions determined by one, two, three, four, . . ., $n + 1$ points, and having zero, one, two, three, . . ., n dimensions, i.e. an r-flat is determined by $r + 1$ points, and every p-flat ($p < r$) which is determined by $p + 1$ of these points lies entirely in the r-flat. We shall suppose that the n-flat, or space of n dimensions, contains all the points. A p-flat, or hyperplane of p dimensions will be denoted by S_p. A flat space is also called a *linear space*.

11. Independent Points. If $p + 1$ points uniquely determine a p-flat they must not be contained in the same $(p - 1)$-flat. Also no r of them ($r \gtrless p$) must be contained in the same $(r - 2)$-flat, for this $(r - 2)$-flat, which is determined by $r - 1$ points, together with the remaining $p + 1 - r$ points, would determine a $(p - 1)$-flat. We shall call a system of $p + 1$ points, no r of which lie in the same $(r - 2)$-flat, a system of *linearly independent* points. Any $p + 1$ points of a p-flat, if they are linearly independent, can be chosen to determine the p-flat.

12. Consider a p-flat and a q-flat, which are determined respectively by $p + 1$ and $q + 1$ points. If they have no point in common we have $p + q + 2$ independent points which determine a $(p + q + 1)$-flat. Hence *a p-flat and a q-flat taken arbitrarily lie in the same $(p + q + 1)$-flat.*

If $p + q + 1$ is greater than n the two flats will have a region in common. Let this region be of dimensions r. In this region we may take $r + 1$ independent points; to determine the p-flat we require $p - r$ additional points, and to determine the q-flat $q - r$ additional points, i.e. altogether

$(r + 1) + (p - r) + (q - r) = p + q - r + 1$, and these determine a $(p + q - r)$-flat. Hence

A p-flat and a q-flat which have in common an r-flat are both contained in a $(p + q - r)$-flat; if they have no point in common they are both contained in a $(p + q + 1)$-flat.

A p-flat and a q-flat, which are both contained in an n-flat, have in common a $(p + q - n)$-flat, provided $p + q > n - 1$. If $p + q < n$ they have no point in common. When two flats do not intersect we may say that they intersect in a (-1)-flat.

13. Degrees of Freedom of Linear Spaces. A p-flat requires $p + 1$ points to determine it, and each point requires n conditions to determine it in space of n dimensions. But we have in the choice of each point p degrees of freedom. Hence the number of conditions required to determine the p-flat in space of n dimensions is $(p+1)(n-p)$, i.e. *the number of degrees of freedom of a p-flat in an n-flat is* $(p + 1)(n - p)$. This is called the *constant-number* of the p-flat.

This result may be proved otherwise, thus. Take any $p + 1$ fixed $(n - p)$-flats. The p-flat cuts each of these in a point, and these $p + 1$ points determine the p-flat. Each point has $n - p$ degrees of freedom in its $(n - p)$-flat. Hence the total number of degrees of freedom of the p-flat is $(p+1)(n-p)$.

If the p-flat has r points fixed, $p + 1 - r$ points are still required to determine it, hence the number of degrees of freedom of the p-flat is $(n - p)(p - r + 1)$; hence also *the number of degrees of freedom of a p-flat lying in a given n-flat and passing through a given r-flat is* $(n - p)(p - r)$.

14. The degree of incidence of a p-flat and a q-flat can be represented by a fraction. Let $p > q$. Complete incidence, when the p-flat contains the q-flat, can be represented by 1; skewness, when they have no point in common, by 0. If they have in common an r-flat, the degree of incidence can be represented by the fraction $(r + 1) / (q + 1)$.

15. Duality or Reciprocity. A p-flat and an $(n - p - 1)$-flat in S_n have the same constant-number $(p + 1)(n - p)$. A $(1, 1)$ correspondence between points and $(n - 1)$-flats can be established in various ways, so that to the line joining two points P, Q corresponds the $(n - 2)$-flat of intersection of the two corresponding $(n - 1)$-flats. If three points are collinear their corresponding $(n - 1)$-flats pass through the same $(n - 2)$-flat. To a $(p - 1)$-flat, which is determined by p given points, corresponds the $(n - p)$-flat common to the $(n - 1)$-flats which correspond to the p points.

16. Number of Conditions Required for a given Degree of Incidence. In S_n a p-flat has $(n - p)(p + 1)$ degrees of freedom, but if it passes through a given r-flat it has only $(n - p)(p - r)$ degrees of freedom. Hence *the number of conditions that a p-flat in S_n should pass through a given r-flat* $(n > p > r)$ is $(n - p)(r + 1)$.

If the r-flat is free to move in a given q-flat, it has $(r + 1)(q - r)$ degrees of freedom. Hence *the number of conditions that a p-flat and a q-flat in S_n should intersect in an r-flat is* $(r + 1)(n - p - q + r)$. This implies that $p + q \gtrless n + r$. If $p + q > n + r$ they intersect in a region of dimensions $p + q - n$, which is greater than r.

17. Incidence of a Linear Space with two or more Linear Spaces. Enumerative Geometry. In S_3 a line has 4 degrees of freedom (constant-number 4), and the number of conditions that it should intersect another line is 1. Hence a line which cuts a fixed line has 3 degrees of freedom, and the whole system of lines all cutting a fixed line forms a three-dimensional assemblage which is a particular case of a *congruence*; the system of lines cutting two fixed lines forms a two-dimensional assemblage, a particular case of a *complex*; and the system of lines cutting three fixed lines forms a one-dimensional assemblage, a *line-series.* A line which is required to cut four given lines is deprived of all freedom. The line is not, however, uniquely determined. An important problem, which belongs to a branch of mathematics called *enumerative geometry*, is to determine the number of linear spaces which satisfy given conditions, in number equal to the constant-number of the linear space.

This problem can sometimes be solved directly by simple geometrical considerations. As an example let us find the number of lines in S_5 (constant-number 8) which pass through a given point O (4 conditions) and cut two given planes a, b (2 conditions for each intersection). The required line must lie in each of the 3-flats (Oa) and (Ob), hence it is *uniquely* determined as the intersection of these two 3-flats.

18. Principle of Specialisation. The determination of the number of lines which cut four lines in S_3 is not so simple, and we apply a very useful principle, called by Schubert the "conservation of number" (Erhaltung der Anzahl), or "principle of specialisation." The principle is that the number of elements determined will be the same if the determining figures are specialised, provided the number does not thereby become infinite. In simple cases it is equivalent to the statement in algebra that the number of roots of an algebraic equation is not altered if the coefficients are specialised in any way, unless, indeed, the equation becomes an identity.

As an example of the application of this principle let us complete the problem to determine how many lines in S_3 cut four given lines. Let the four lines intersect in pairs: a and b in P, c and d in Q. A line which cuts both a and b must either pass through P or lie in the plane (ab); and as P must not lie in the plane (cd), nor Q in the plane (ab), for in either of these cases there would be an unlimited number of lines cutting all four lines, the required line must either pass through both P and Q, or lie in both of the planes (ab) and (cd). Hence there are *two* lines satisfying the given conditions.

19. The general enumerative problem relating to linear spaces is: to find the number of p-flats which have incidence of specified degrees with a given set of linear spaces, the number of assigned conditions being equal to the constant-number of the p-flat.

Let $(n:p, q:r)$ represent the condition that a p-flat and a q-flat in S_n should intersect in an r-flat. If P and Q represent any conditions, the product PQ is taken to mean that both conditions must be simultaneously satisfied; the sum P + Q that one or other of the conditions is satisfied. Thus $(3:2, 0:0)(3:2, 1:1)$ means the condition that a plane in S_3 should contain a given point and a given line; $(3:1, 1:0)^4$ means the condition that a line in S_3 should cut four given lines.

The number of simple conditions involved in $(n:p, q:r)$ may be denoted by $C(n:p, q:r)$, and we have proved that

$$C(n:p, q:r) = (r+1)(n-p-q+r).$$

The constant-number of a p-flat in S_n is equal to

$$C(n:p, p:p) = (p+1)(n-p).$$

We may also represent the number of elements determined by a given condition by prefixing N to the symbol of the condition. Thus $N(3 : 1, 1 : 0)^4 = 2$.

20. Duality in Enumerative Geometry. The duality between the p-flat and the $(n - p - 1)$-flat extends to enumerative problems. Not only are the constant-numbers equal, viz. $(p + 1)(n - p)$, but we have also, as is easily verified,

$$C(n : p, q : r) = C(n : n - p - 1, n - q - 1 : n - p - q + r - 1);$$

and the number of p-flats determined by $(p + 1)(n - p)$ simple conditions is equal to the number of $(n - p - 1)$-flats determined by the corresponding reciprocal conditions. Thus in S_4 there are two lines which lie in a given 3-flat and cut four given planes, for the planes cut the 3-flat in lines. The reciprocal statement is: there are two planes which pass through a given point and cut four given lines; i.e.

$$N(4 : 1, 3 : 1)(4 : 1, 2 : 0)^4 = N(4 : 2, 0 : 0)(4 : 2, 1 : 0)^4.$$

21. In two dimensions the only conditions are

$$C(2 : 0, 0 : 0) = 2 = C(2 : 1, 1 : 1),$$
$$C(2 : 0, 1 : 0) = 1 = C(2 : 1, 0 : 0),$$

i.e. a point coincident with a given point, a line coincident with a given line; and a point incident with a given line, a line incident with a given point. The constant-number K of a point or a line is 2, and we have only two enumerative results with their duals, viz. :—

$N(2 : 0, 0 : 0) = 1$, *i.e.* one point is coincident with a given point,
$N(2 : 0, 1 : 0)^2 = 1$, *i.e.* one point is incident with two given lines,
$N(2 : 1, 1 : 1) = 1$, *i.e.* one line is coincident with a given line,
$N(2 : 1, 0 : 0)^2 = 1$, *i.e.* one line is incident with two given points.

The results may be exhibited more compactly in tabular form. For a specified linear space let q_r denote that it cuts a given q-flat in an r-flat. Then for S_3 all the results are represented in the following table. Point and plane, being reciprocals, are grouped together.

Point Plane	0_0 2_2	1_0 1_1	2_0 0_0	K = 3
C	3	2	1	N
	1	.	.	1
	.	1	1	1
	.	.	3	1

Line	0_0	1_0	1_1	2_1	K = 4
C	2	1	4	2	N
	2	.	.	.	1
	1	2	.	.	1
	1	.	.	1	0
	.	4	.	.	2
	.	2	.	1	1
	.	.	1	.	1
	.	.	.	2	1

C is the number of simple conditions, N is the number of elements determined. In each row the sum of the numbers, each multiplied by the corresponding value of C, is equal to the constant-number of the element K.

22. Incident Spaces in Four Dimensions. In general many of the combinations are either trivial, or belong to a space of lower dimensions, e.g. in S_4 the statement, that one line passes through a given point and intersects three given lines, is the reciprocal of the statement that one plane lies in a given 3-flat and cuts three given lines (the plane is in fact determined by the three points in which the given lines cut the 3-flat).

The following is the complete table of results for S_4, most of which the reader will have no difficulty in verifying. Results marked "trivial" are the fundamental determinations of an element or its reciprocal, e.g. a plane determined by three points. "S_3" indicates that the result belongs to space of three dimensions.

Point 3-flat	0_0 3_3	1_0 2_2	2_0 1_1	3_0 0_0	K = 4	
C	4	3	2	1	N	
	1	.	.	.	1	all trivial
	.	1	.	1	1	
	.	.	2	.	1	
	.	.	1	2	1	
	.	.	.	4	1	

Line Plane	0_0 3_2	1_0 2_1	2_0 1_0	1_1 2_2	2_1 1_1	3_1 0_0	K = 6	
C	3	2	1	6	4	2	N	
	2	.	.	.	.	.	1	trivial
	1	1	1	.	.	.	1	S_3
	1	.	3	.	.	.	1	S_3
	1	.	1	.	.	1	0	
	.	3	.	.	.	.	1	
	.	2	2	.	.	.	2	
	.	2	.	.	.	1	1	S_3
	.	1	4	.	.	.	3	
	.	1	2	.	.	1	1	S_3
	.	1	.	.	1	.	0	
	.	1	.	.	.	2	0	
	.	.	6	.	.	.	5	
	.	.	4	.	.	1	2	S_3
	.	.	2	.	1	.	1	trivial, S_2
	.	.	2	.	.	2	1	trivial, S_2
	.	.	.	1	.	.	1	trivial
	.	.	.	.	1	1	1	trivial
	.	.	.	.	.	3	1	trivial

The most important of these conditions which determine a line are: 1_0^3, $1_0^2 2_0^2$, $1_0 2_0^4$, and 2_0^6. These may be investigated as follows.

1_0^3. Let the three given lines be a, b, c. A line which cuts both a and b lies in the 3-flat (ab). The three 3-flats (bc), (ca), (ab) have a unique line in common. Hence $N1_0^3 = 1$.

$1_0^2 2_0^2$. The two given lines determine a 3-flat which cuts the two given planes in two lines. These four lines have two common transversals. Hence $N1_0^2 2_0^2 = 2$.

$1_0 2_0^4$. Denote the given line by a and the given planes by α, β, γ, δ. Using the principle of specialisation, let two of the given planes α, β intersect in a line l. A line which cuts both α and β either cuts l or lies in the 3-flat $(\alpha\beta)$. In the first case the condition becomes $1_0^2 2_0^2$ and two lines are determined; in the second case it reduces to $0_0 1_0^2$ in S_3, and one line is determined. Hence $N1_0 2_0^4 = 3$.

2_0^6. Let the six given planes intersect in pairs in straight lines: α, α' in a; β, β' in b; γ, γ' in c. A line which meets both α and α' must either cut a or lie in the 3-flat

$(\alpha\alpha')$. Hence the required line either (1) cuts three given lines, (2) cuts two given lines and lies in a given 3-flat (three different ways), (3) cuts a given line and lies in two given 3-flats (also three different ways), or (4) lies in three given 3-flats. Under the first category one line is determined; under the second, three; no line exists under the third category, since the given line does not in general cut the plane common to the two 3-flats, and if it did, an infinity of lines would be possible; lastly, under the fourth category one line is determined. Hence $N2_0^6 = 5$.

23. Incident Spaces in Five Dimensions. In five dimensions we have the reciprocal pairs: point and 4-flat, line and 3-flat, plane (self-reciprocal). Omitting the cases of point and 4-flat, which are trivial, the following are the complete results (see p. 16).

With the exception of Nos. 16, 20, 21, 23, 27 and 34 these can all be proved directly or deduced from previous results in S_4 or S_3. 20 and 23 can be solved by letting two of the planes intersect in a point; 16, 21, 27 and 34 by letting two of the 3-flats intersect in a plane.

Line	0_0	1_0	2_0	3_0	1_1	2_1	3_1	4_1	K = 8
3-flat	4_3	3_2	2_1	1_0	3_3	2_2	1_1	0_0	
C	4	3	2	1	8	6	4	2	N
1	2	.	.	.	.	.	.	.	1
2	1	1	.	1	.	.	.	.	1
3	1	.	2	.	.	.	.	.	1
4	1	.	1	2	.	.	.	.	1
5	1	.	1	.	.	.	.	1	0
6	1	.	.	4	.	.	.	.	1
7	1	.	.	2	.	.	.	1	0
8	1	.	.	.	.	.	1	.	0
9	1	.	.	.	.	.	.	2	0
10	.	2	1	.	.	.	.	.	1
11	.	2	.	2	.	.	.	.	2
12	.	2	.	.	.	.	.	1	1
13	.	1	2	1	.	.	.	.	2
14	.	1	1	3	.	.	.	.	3
15	.	1	1	1	.	.	.	1	1
16	.	1	.	5	.	.	.	.	4
17	.	1	.	3	.	.	.	1	1
18	.	1	.	1	.	.	1	.	0
19	.	1	.	1	.	.	.	2	0
20	.	.	4	.	.	.	.	.	3
21	.	.	3	2	.	.	.	.	4
22	.	.	3	.	.	.	.	1	1
23	.	.	2	4	.	.	.	.	6
24	.	.	2	2	.	.	.	1	2
25	.	.	2	.	.	.	1	.	1
26	.	.	2	.	.	.	.	2	1
27	.	.	1	6	.	.	.	.	9
28	.	.	1	4	.	.	.	1	3
29	.	.	1	2	.	.	1	.	1
30	.	.	1	2	.	.	.	2	1
31	.	.	1	.	.	1	.	.	0
32	.	.	1	.	.	.	1	1	0
33	.	.	1	.	.	.	.	3	0
34	.	.	.	8	.	.	.	.	14
35	.	.	.	6	.	.	.	1	5
36	.	.	.	4	.	.	1	.	2
37	.	.	.	4	.	.	.	2	2
38	.	.	.	2	.	1	.	.	1
39	.	.	.	2	.	.	1	1	1
40	.	.	.	2	.	.	.	3	1
41	.	.	.	.	1	.	.	.	1
42	.	.	.	.	.	1	.	1	1
43	.	.	.	.	.	.	2	.	1
44	.	.	.	.	.	.	1	2	1
45	.	.	.	.	.	.	.	4	1

Plane	0_0	1_0	2_0	1_1	2_1	3_1	2_2	3_2	4_2	K = 9
	4_2	3_1	2_0	3_2	2_1	1_0	2_2	1_1	0_0	
C	3	2	1	6	4	2	9	6	3	N
1, 1′	3	.	.	.	.	.	.	.	.	1
2, 2′	2	1	1	.	.	.	.	.	.	1
3, 3′	2	.	3	.	.	.	.	.	.	1
4, 4′	2	.	1	.	.	1	.	.	.	0
5, 5′	2	.	.	.	.	.	.	.	1	0
6, 6′	1	3	.	.	.	.	.	.	.	1
7, 7′	1	2	2	.	.	.	.	.	.	2
8, 8′	1	2	.	.	.	1	.	.	.	1
9, 9′	1	1	4	.	.	.	.	.	.	3
10, 10′	1	1	2	.	.	1	.	.	.	1
11, 11′	1	1	1	.	.	.	.	.	1	0
12, 12′	1	1	.	.	1	.	.	.	.	0
13, 13′	1	1	.	.	.	2	.	.	.	0
14, 14′	1	.	6	.	.	.	.	.	.	5
15, 15′	1	.	4	.	.	1	.	.	.	2
16,	1	.	3	.	.	.	.	.	1	0
17, 17′	1	.	2	.	1	.	.	.	.	1
18, 18′	1	.	2	.	.	2	.	.	.	1
19, 19′	1	.	.	1	.	.	.	.	.	1
20, 20′	1	.	.	.	1	1	.	.	.	1
21, 21′	1	.	.	.	.	3	.	.	.	1
22, 22′	1	.	.	.	.	.	.	1	.	0
23, 23′	.	4	1	.	.	.	.	.	.	3
24, 24′	.	3	3	.	.	.	.	.	.	6
25, 25′	.	3	1	.	.	1	.	.	.	3
26, 26′	.	2	5	.	.	.	.	.	.	11
27, 27′	.	2	3	.	.	1	.	.	.	5
28, 28′	.	2	1	.	1	.	.	.	.	1
29	.	2	1	.	.	2	.	.	.	2
30, 30′	.	1	7	.	.	.	.	.	.	21
31	.	1	5	.	.	1	.	.	.	10
32, 32′	.	1	3	.	1	.	.	.	.	3
33, 33′	.	1	1	1	.	.	.	.	.	1
34	.	1	1	.	1	1	.	.	.	2
35, 35′	.	1	1	.	.	.	.	1	.	0
36	.	.	9	.	.	.	.	.	.	42
37	.	.	5	.	1	.	.	.	.	6
38, 38′	.	.	3	1	.	.	.	.	.	1
39	.	.	1	.	2	.	.	.	.	0
40	.	.	.	.	.	.	1	.	.	1

This table includes 72 cases, 8 of which are self-reciprocal. Thus No. 1 is 0_0^3 and No. 1′, its reciprocal, is 4_2^3. Nos. 1 to

22, 33, 35, 38 and 40, can be proved directly or deduced from results in S_3 or S_4. For the others we may use the principle of conservation of number. If the condition involves the factor 2_0^2, let the two planes α, β cut in a line l. Then the required plane either cuts l, or cuts the 3-flat $(\alpha\beta)$ in a line. Hence

$$2_0^2 = 1_0 + 3_1.$$

This leads to a number of equations, such as

$$\begin{aligned} N(1_0^3 2_0^3) &= N(1_0^4 2_0) + N(1_0^3 2_0 3_1), \\ N(1_0 2_0^7) &= N(1_0^2 2_0^5) + N(1_0 2_0^5 3_1) \\ &= N(1_0^3 2_0^3) + 2N(1_0^2 2_0^3 3_1) + N(1_0 2_0^3 3_1^2) \\ &= N(1_0^4 2_0) + 3N(1_0^3 2_0 3_1) + 3N(1_0^2 2_0 3_1^2) + N(1_0 2_0 3_1^3), \end{aligned}$$

and so on. Next, if the condition involves the factor $2_0^2 3_1$, let the two planes a, b meet the 3-flat α in lines l, m. Then $(a\alpha)$ and $(b\alpha)$ are 4-flats. The required plane then either lies in $(a\alpha)$ and cuts b, or lies in $(b\alpha)$ and cuts a, or cuts both l and m. Hence

$$2_0^2 3_1 = 2(2_0 4_2) + 1_0^2.$$

The equations derived from this, together with the previous ones, will be found to determine all but Nos. 28, 32 and 37, which involve the factor 2_1. If we assume the relation

$$1_0 2_0 = x(0_0) + y(2_0^3) + z(2_0 3_1) + w(4_2),$$

where x, y, z, w are certain unknown numbers, we have, by applying this to $0_0^2 1_0 2_0$, $0_0 1_0 2_0^2 3_1$ and $1_0^4 2_0$, the relations $1 = x + y$, $1 = 2y + z$, and $3 = x + 6y + 3z + w$, giving $y = 1 - x$, $z = 2x - 1$, $w = -x$. Then applying it to $1_0 2_0^3 2_1$ we find $N(2_0^5 2_1) = 2N(1_0 2_0^3 2_1)$; and applying it to $1_0^2 2_0 2_1$ and $1_0 2_0 2_1 3_1$ we find $N(1_0^2 2_0 2_1) = 1$ and $N(1_0 2_0^3 2_1) = 3$. The x here plays the part of a catalytic agent.

24. We shall conclude this section with a few general results.

In S_n the constant-number of a line is $2(n - 1)$, and one condition is required in order that a line should cut a given $(n - 2)$-flat, hence a finite number of lines in S_n cut $2(n - 1)$ linear spaces of $n - 2$ dimensions. It was proved by W. F. Meyer that the number of lines is $\frac{1}{n-1}\,{}_{2n-2}C_n = \frac{(2n-2)!}{n!\,(n-1)!}$. A more general result, which includes this as a particular case, was proved by Schubert, viz. *the number of p-flats in S_n which cut $n - p$ given $(n - 2p - 1)$-flats is*

$$\frac{1!\,2!\,3!\,\ldots\,p!\,\{(p+1)(n-p)\}!}{n!\,(n-1)!\,(n-2)!\,\ldots\,(n-p)!}.$$

To find the number of lines in S_{2k-1} which cut 4 linear spaces of $(k-1)$ dimensions. The constant-number of a line in S_{2k-1} is $4(k - 1)$, and the number of conditions that it should cut a $(k - 1)$-flat is $k - 1$. Hence there is a finite number. Let the $(k - 1)$-flats intersect in pairs in points: α, β in P, and γ, δ in Q. A line which cuts α and β either passes through P or lies in the $(2k - 2)$-flat $(\alpha\beta)$. We have then (1) one line through P and Q, (2) lines through P and lying in the $(2k - 2)$-flat [none], (3) lines in the $(2k - 3)$-flat common to $(\alpha\beta)$ and $(\gamma\delta)$, and cutting the four $(k - 2)$-flats in which α, β are cut by $(\gamma\delta)$, and γ, δ by $(\alpha\beta)$. Hence if $f(k)$ is the number of lines which cut four $(k - 1)$-flats in S_{2k-1},

$$f(k) = f(k-1) + 1.$$

But, for $k = 2$, we have 2 lines in S_3 cutting 4 lines, i.e. $f(2) = 2$. Hence $f(k) = k$.

Ex. (1) Prove that there is one 3-flat in S_{2q} cutting three q-flats in lines.

(2) One plane in S_{2q} cutting three q-flats in lines.

(3) One plane in S_{4q+2} cutting four q-flats in points.

(4) One plane in S_{8q+2} cutting four $(5q + 1)$-flats in lines.

25. Motion and Congruence. Strictly speaking the idea of motion is foreign to geometry. When we speak of a point moving along a line we really mean that we are focussing our attention successively on a series of points in the line. When we speak of a figure being displaced in space we are really transferring our attention from one figure to another (congruent) figure; or we are considering a transformation in which certain relations, distances and angles, are invariant. The idea of motion thus involves sequence or *order*, and *congruence*. For these there are two corresponding groups of axioms. As these axioms deal almost exclusively with plane geometry, and the new ideas involving congruence in higher space can be dealt with by definitions, we may simply refer for a discussion of the axioms to Hilbert's "Foundations of Geometry" or to the more recent "Foundations of Euclidean Geometry," by Forder.

26. Order. As regards order in euclidean geometry, a point divides a line into two parts, but does not of course divide a plane; a line divides a plane, but does not divide space; a plane divides space, but does not divide a 4-flat, for in S_4 the line joining two points does not in general intersect a given plane: and so on.

If A_1 and A_2 are given points on a line they determine a *segment* which consists of all points P such that A_1PA_2 are in this order. A_1 and A_2 divide the line into three parts, for which we have respectively the orders PA_1A_2, A_1PA_2, A_1A_2P.

If A_1, A_2, A_3 are three non-collinear points in a plane, they determine three lines which form a triangle. The segments A_2A_3, etc., are the sides or edges, and if A_{23} denotes any point on the segment A_2A_3 the interior of the triangle consists of all points P such that A_1PA_{23} are in this order. The three lines divide the plane into seven regions, consisting of points characterised by the following orders:

(1) the interior of the triangle: A_1PA_{23} or A_2PA_{31} or A_3PA_{12},
(2, 3, 4) three regions on the faces: $A_1A_{23}P$, $A_2A_{31}P$, $A_3A_{12}P$,
(5, 6, 7) three regions on the vertices: PA_1A_{23}, PA_2A_{31}, PA_3A_{12}.

Similarly four non-coplanar points A_1, A_2, A_3, A_4 determine four planes and six lines forming a tetrahedron. Its faces are the interiors of the triangles $A_2A_3A_4$, etc. The four planes divide space into 15 regions which consist of points P characterised by the following orders, where A_{234} denotes any point on the face $A_2A_3A_4$, and A_{12} any point on the edge A_1A_2:

the interior of the tetrahedron: A_1PA_{234}, etc.,
four regions on the faces: $A_1A_{234}P$, etc.,
four regions on the vertices: $A_{234}A_1P$, etc.,
six regions on the edges: $A_{12}A_{34}P$, etc.

(In explanation of the last statement we note that the plane PA_{12} cuts the line A_3A_4 in a point, and if P is in the region specified this point lies in the segment A_3A_4.)

In space of four dimensions we have similarly five points determining 5 hyperplanes, 10 planes, and 10 edges, forming a four-dimensional *simplex*, and dividing S_4 into 31 regions:

the interior of the simplex:	A_1PA_{2345}, etc.,
5 regions on the 3-dimensional boundaries:	$A_1A_{2345}P$, etc.,
10 regions on the 2-dimensional boundaries:	$A_{12}A_{345}P$, etc.,
10 regions on the edges:	$A_{123}A_{45}P$, etc.,
5 regions on the vertices:	$A_{1234}A_5P$, etc.

The extension to n dimensions is now obvious. The figure formed by $n + 1$ independent points and the lines, planes, etc., directly determined by them is called a *simplex* of n dimensions and will be denoted by $S(n + 1)$. The lines, planes, etc., are called its boundaries of one, two, etc., dimensions; the points are its vertices. An $S(n + 1)$ has ${}_{n+1}C_{r+1}$ boundaries of r dimensions.

REFERENCES

CAYLEY, A. A memoir on abstract geometry. Phil. Trans. **160** (1870), 51-63; Math. Papers, vi. No. 413.

FORDER, H. G. The foundations of euclidean geometry. Cambridge: Univ. Press, 1927.

HILBERT, D. The foundations of geometry. Authorised trans. by E. J. Townsend. Chicago: Open Court, 1902. (Original German edition: Grundlagen der Geometrie, Leipzig, 1899. 6. Aufl. 1923).

MANNING, H. P. Geometry of four dimensions. New York: Macmillan, 1914. (Synthetic method. The introduction contains an interesting short history of the subject.)

MEYER, W. F. Apolarität und rationale Curven. Eine systematische Voruntersuchung zu einer allgemeinen Theorie der linearen Räumen. Tübingen, 1882.

SCHUBERT, H. Die n-dimensionalen Verallgemeinerungen des dreidimensionalen Satzes, dass es zwei Strahlen giebt, welche vier gegebene Strahlen schneiden. Hamburg, Mitt. Math. Ges., No. 4, 1884.

— Kalkül der abzählenden Geometrie. Leipzig, 1879.

SEGRE, C. Mehrdimensionale Räume. Encykl. Math. Wiss. iii. c. 7. 1921.

CHAPTER II

PARALLELS

1. Parallel Lines. In projective geometry two straight lines in a plane always intersect. This is not true in euclidean or hyperbolic geometry.

Let l be a given line (Fig. 4) and O a given point not on l; A any fixed point on l, and P a variable point on l. Let P move always in one direction along l, i.e. we consider a series of points P_1, P_2, . . . in the order AP_1P_2 . . . As P moves along l it may return to its initial position. This occurs in projective geometry, and also in elliptic geometry, in which two lines in a plane always intersect. Otherwise as the segment AP increases without limit, the line OP will tend to a definite limiting position OL. The line OL, which separates the rays through O which intersect AP from those which do not, is said to be *parallel* to AP. If P moves in the opposite sense we get another limiting position OL′, and OL′ is parallel to AP′.

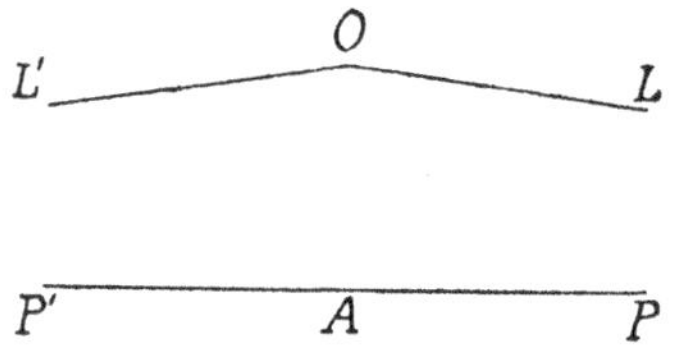

Fig. 4.

In euclidean geometry the two rays OL, OL′ form one and the same line, but in hyperbolic geometry they belong to distinct lines, and we have to distinguish the two directions of parallelism to a given straight line. In projective geometry the foregoing considerations are irrelevant, since projective geometry does not involve the idea of distance.

2. A fundamental theorem is the following:

If AA′ ‖ CC′, and BB′ ‖ CC′, then AA′ ‖ BB′.

(1) Let the three lines be coplanar, and (i) let CC′ lie between AA′ and BB′ (Fig. 5). We may assume that A, C, B

are collinear. Within the angle BAA′ draw any line AP. Since AA′ ∥ CC′, AP cuts CC′ in a point R. Then since RC′ ∥ BB′, PR produced must cut BB′. Also AA′ does not cut BB′, therefore AA′ ∥ BB′.

(ii) In the same figure let AA′ ∥ BB′, and CC′ ∥ BB′. AA′ cannot cut CC′, for it would then cut BB′, but any other line within the angle BAA′ cuts CC′, therefore AA′ ∥ CC′.

(2) Let the three lines be not all in the same plane, and let AA′ ∥ BB′ and CC′ ∥ BB′ (Fig. 6). Take any point P on BB′. Then as P moves along BB′, AP → AA′, and CP → CC′, while P, A, C are always coplanar, therefore AA′ and CC′ are coplanar. Again; if CP is fixed while the plane PAC moves, PA → PB′, and the plane CPA → CPB′. CA, the line of intersection of CPA and C′CAA′, therefore → CC′, and CC′ ∥ AA′.

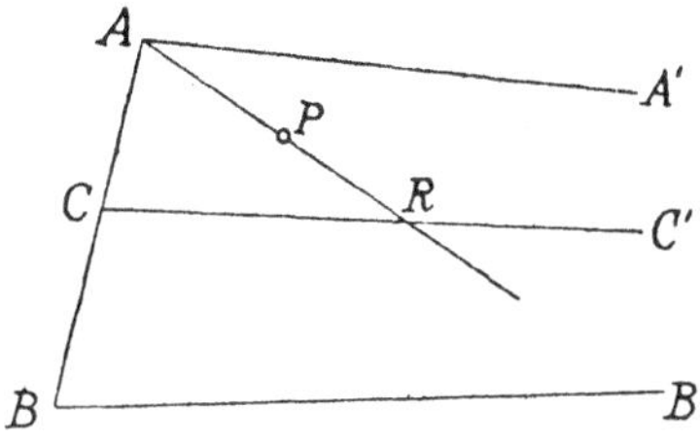

FIG. 5.

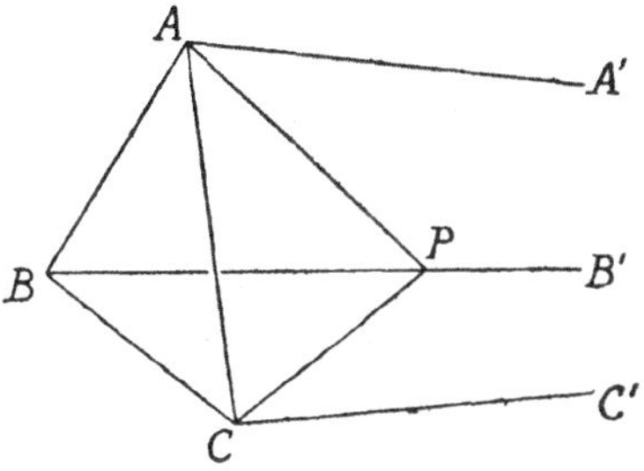

FIG. 6.

3. Parallel Lines and Planes. We shall confine our attention now for the present to euclidean geometry. In a plane two straight lines either intersect in one point or are parallel. In three dimensions two planes which have a point in common intersect in a straight line. If they have no point in common they are said to be parallel. Two parallel planes are cut by a third plane in two straight lines which have no point in common and are therefore parallel. A straight line either cuts a given plane in one point or has no point in common with the plane; in the latter case the straight line is said to be parallel to the plane.

Consider a plane α and a point O not in α. Let β be any plane through O cutting α in a line b. Then through O and in β there is one line parallel to b and therefore parallel to α.

By taking different planes β we get a system of lines through O all parallel to α. Let l, m, n be three such lines. The planes (lm) and (ln) are both parallel to α, and must coincide, for let γ be a plane through O cutting α in c, (lm) in g, and (ln) in h. Then g and h are both parallel to c and therefore (in euclidean geometry) coincide. Hence all the lines through O parallel to α lie in one plane which is parallel to α.

4. **Direction and Orientation.** By an extension of the meanings of the words point and line we can greatly simplify the statements of the relations between parallel lines and planes. A system of concurrent lines in a plane, called a *pencil* of lines, determine a unique element, their common point, and this is determined by any two lines of the system; just as a system or range of collinear points determine a unique element, their common line, which is determined by any two points of the system. *Pencil of lines* is said to be *reciprocal* or *dual* to *range of points*. In three dimensions a system of concurrent lines, or *bundle* of lines, is of two dimensions, and is reciprocal to the *plane field of points*. A system of straight lines all parallel to the same straight line also determine a unique element, a *direction*, which is uniquely determined by any two of the lines. A direction together with a point uniquely determine a straight line. A direction together with two points uniquely determine a plane, for one of the points together with the given direction determine a line, and this with the other point determine a plane. Similarly two directions together with a point uniquely determine a plane. Thus a direction may take the place of a point in determining lines or planes.

Two directions by themselves do not determine a plane, but only the *orientation* of a plane; and we may then say an orientation together with a point uniquely determine a plane. Also if two planes have their orientations given, their line of intersection has a fixed direction, thus two orientations uniquely determine a direction. An orientation may thus take the place of a line in determining planes and points. As, however, two lines only determine a point when they are coplanar, while two orientations (in space of three dimensions) always determine a direction, orientations appear to correspond to coplanar lines.

5. **Points at Infinity.** To bring out more explicitly the

connection between direction and point, orientation and line, we give them the names "point at infinity" and "line at infinity" respectively, and we shall be justified in extending to them the language used for points and lines. We speak of the "point at infinity on a line" for the direction of a line, the "line at infinity on a plane" for the orientation of a plane. A "point at infinity on a plane" is the direction of some line in that plane. The orientation determined by two directions is said to contain the two directions; thus a line at infinity is determined by two points at infinity; and, since the directions of any two lines of a given plane determine the orientation of the plane, it follows that any two points at infinity on a plane determine or lie on the same line at infinity.

We may now draw correct logical conclusions by picturing a line at infinity as a line whose elements are points at infinity. Parallel lines can then be pictured as lines which meet in a point at infinity. In a given plane we picture the line at infinity as a special line l_∞ in the plane, and the condition for parallelism of two lines a, b in the plane is expressed by the concurrency of a, b and l_∞. Since (in three dimensions) two lines at infinity always determine a point at infinity, we have to picture the assemblage of all lines at infinity as lines in one common plane. The assemblage of all orientations in space of three dimensions is called the *plane at infinity* in this space. We may denote this by π_∞. The point at infinity on any line, and the line at infinity on any plane, is its intersection with the plane at infinity. Two lines are parallel when they intersect on the plane at infinity; a line is parallel to a plane when the point at infinity on the line lies on the line at infinity in the plane; two planes are parallel when their lines at infinity coincide. Two arbitrary planes, α, β have in common one point at infinity, the common point of the three planes α, β, π_∞, or the common point of their lines at infinity. In ordinary language, if α is not parallel to β, it is always possible to get two series of parallel lines, one in α and one in β, viz., the lines which are parallel to the line of intersection of α and β.

6. Parallel and Half-parallel Planes. We shall consider now parallelism in euclidean space of four dimensions. Two hyperplanes which have a point in common intersect in a plane.

If they have no point in common they are said to be parallel. Two parallel hyperplanes are cut by a third hyperplane in two planes which have no point in common and are therefore parallel in the ordinary sense. Two planes in general intersect in just one point. We have defined parallel planes when they lie in space of three dimensions and meet in a line at infinity. But in space of four dimensions two planes may have another sort of parallelism when they do not lie in the same hyperplane and have no point in common. The term "half-parallel" or "semi-parallel" has been used to describe this.

7. The Hyperplane at Infinity. Two hyperplanes each parallel to a third are parallel to one another. We shall say that two parallel hyperplanes determine a *stratification*, just as two planes in three dimensions determine an orientation. Two stratifications determine an orientation; three stratifications determine a direction. All the orientations in a given hyperplane belong to the stratification of that hyperplane, so that the stratification of a hyperplane is identified with the plane at infinity in that hyperplane. Since two planes at infinity always determine a line at infinity we have to picture the assemblage of all planes at infinity as planes in one common hyperplane, so we call the assemblage of all planes at infinity the *hyperplane at infinity* in space of four dimensions.

We can now perceive the distinction between parallel and semi-parallel planes. Parallel planes have in common a straight line at infinity, semi-parallel planes have in common only a point at infinity. In ordinary language, parallel planes have the same orientation and an infinity of common directions; to every line in the one plane there is a system of lines parallel to it in the other. Semi-parallel planes have only one common direction; there is only one system of parallel lines in the one plane which are parallel to lines in the other.

8. Degrees of Parallelism. We can now extend the idea of parallelism to space of n dimensions. In euclidean space of n dimensions there is a unique $(n - 1)$-flat at infinity. Two $(n - 1)$-flats are parallel when their common $(n - 2)$-flat is an $(n - 2)$-flat at infinity, i.e. an element of the $(n - 1)$-flat at infinity. Two planes, in addition to the possibility of being parallel or semi-parallel, may have no point at all in common and have no common containing hyperplane; they are then skew. Two 3-flats which have no point in common (i), if they lie in the same 4-flat, have a plane at infinity in common; (ii), if they lie only in the same 5-flat, they have a line at infinity in common; (iii), if they lie only in the same 6-flat, they have a point at infinity in common; and (iv), if they do not lie in the same 6-flat, they are skew. In the three cases (i), (ii), (iii) the two 3-flats are said to be (i) completely parallel, (ii) two-thirds parallel, (iii) one-third parallel.

In general, a p-flat and a q-flat ($p \leqq q$), which are both contained in the same $(p + q - r)$-flat, and would therefore in general intersect in an r-flat, but have no point in common, are said to be $(r + 1)/q$ parallel, and have in common an r-flat at infinity. It is to be noted that if a p-flat and a q-flat are contained in the same $(p+q-r)$-flat, they have at least a common $(r - 1)$-flat at infinity, for, if they intersect, their common r - flat contains a unique $(r - 1)$-flat at infinity. Thus, in order that two flats should be parallel it is not sufficient that they should have a certain dimensionality of common points at infinity; they must also have no finite point in common.*

9. Complete parallelism may occur in space of any number of dimensions. Partial parallelism requires a certain minimum dimension. Thus half-parallelism only appears for the first time in S_4 when two planes are half-parallel. One-third parallelism does not appear in space of lower dimensions than six. In general parallelism of order $(r + 1)/q$, where $r + 1$ is prime to q, requires space of at least $2q - r$ dimensions, as it implies the existence of a p-flat and a q-flat contained in the same $(p + q - r)$-flat, and $p \leqq q$.

10. Sections of two Parallel or Intersecting Spaces by a Third Space. If S_p and S_q ($p \leqq q$) are completely parallel, so that the region at infinity $S_{(p-1)\infty}$ on S_p contains $S_{(q-1)\infty}$, S_p and S_q lie in the same $(p + 1)$-flat S_{p+1}, which is determined by $S_{(p-1)\infty}$ and two finite points, one on S_p and one on

* Schoute leaves out the condition of having no finite point in common, and defines a p-flat and a q-flat ($p \leqq q$) as $(r + 1)/q$ parallel when they have in common an r-flat at infinity. According to this, two intersecting planes in three dimensions would be half-parallel since they have in common a point at infinity, viz. the point at infinity on their line of intersection.

S_q. An r-flat ($r > p - q + 1$) in S_{p+1} cuts S_p in an $(r - 1)$-flat S_{r-1} and S_q in a $(q + r - p - 1)$-flat $S_{q+r-p-1}$, and it cuts their common region at infinity in a $(q + r - p - 2)$-flat which is the whole of the region at infinity on $S_{q+r-p-1}$. Hence the two sections are completely parallel.

If S_p and S_q are parallel of degree $(r + 1)/q$ so that they have in common an r-flat at infinity, they lie in the same $(p + q - r)$-flat, which is determined by $r + 1$ common points at infinity, $p - r$ finite points on S_p, and $q - r$ finite points on S_q. An s-flat in this S_{p+q-r} cuts S_p in an $(s+r-q)$-flat and S_q in an $(s + r - p)$-flat, and their common region at infinity in an $(s + 2r - p - q)$-flat. The sections are therefore in general parallel of degree $(s + 2r - p - q + 1)/(s + r - p)$. This fraction $= 1$ if $r = q - 1$.

If S_p and S_q intersect in an S_r and therefore lie in the same S_{p+q-r} they have in common an $S_{(r-1)\infty}$. Let an s-flat S_s parallel to S_r of degree $(m + 1)/r$ ($s \leqq r - 1$) cut S_p and S_q in S_{s+r-q} and S_{s+r-p}. These two sections have then an $S_{m\infty}$ in common and are parallel of degree $(m+1)/(s + r - p)$.

11. The Parallelotope. In n dimensions the analogue of the parallelogram and the parallelepiped is a figure bounded by pairs of parallel $(n - 1)$-flats. Consider a simplex with a vertex O and n edges through O: OA_1, OA_2, . . . , OA_n, no r of which lie in the same $(r - 1)$-flat. The $n - 1$ edges OA_2, . . ., OA_n determine an $(n - 1)$-flat, and through A_1 there is just one $(n - 1)$-flat parallel to this. Constructing the $(n - 1)$-flats, one through each of the n vertices A_1, . . ., A_n parallel to the opposite face of the simplex, we form a figure bounded by $2n$ $(n - 1)$-flats, which is called a *parallelotope.** Each $(n - 1)$-flat is cut by each of the $(n - 1)$-flats except the one which is parallel to it in an $(n - 2)$-flat, and these $2n - 2$ $(n - 2)$-flats are parallel in pairs and form a parallelotope of $n - 1$ dimensions. Each of these again is bounded by parallelotopes of $n - 2$ dimensions, and so on.

The parallelotope may be generated also by successive motions in one, two, three, . . . dimensions. Thus a point

* The suffix *tope* which occurs in several of the terms used in n-dimensional geometry—polytope, simplotope, orthotope, etc.—is from the Greek τόπος = a place or region.

moving in a straight line through a distance a_1 generates a line-segment. This segment moving parallel to itself, so that all its points generate parallel line-segments of length a_2, generates a parallelogram with two edges of length a_1 and two edges of length a_2. The parallelogram by parallel motion similarly generates a parallelepiped, and so on.

Let N_r denote the number of r-dimensional boundaries of a parallelotope of n dimensions, N'_r the corresponding number for a parallelotope of $n - 1$ dimensions; and consider the expression

$$N_0 + N_1x + N_2x^2 + \ldots + N_{n-1}x^{n-1} + N_nx^n,$$

N_n being equal to 1 since the parallelotope itself is the only n-dimensional boundary. The N_r boundaries of r dimensions of the n-dimensional parallelotope are produced by the N'_{r-1} boundaries of $r-1$ dimensions of the $(n-1)$-dimensional parallelotope together with its N'_r boundaries of r dimensions in their initial and final positions. Hence $N_r = N'_{r-1} + 2N'_r$. We have therefore

$$N_0 + N_1x + \ldots + N_nx^n = (N'_0 + N'_1x + \ldots + N'_nx^n)(2+x),$$

and hence by induction

$$N_0 + N_1x + \ldots + N_nx^n = (2 + x)^n,$$

i.e.

$$N_r = {}_nC_r \,.\, 2^{n-r}.$$

The number of r-dimensional boundaries which pass through one vertex is ${}_nC_r$, and all the r-dimensional boundaries fall into ${}_nC_r$ groups, each containing 2^{n-r} completely parallel r-flats.

REFERENCE

SCHOUTE, P. H. Mehrdimensionale Geometrie. 1. Teil: Die linearen Räume. Leipzig: Göschen, 1902.

CHAPTER III

PERPENDICULARITY

1. Line Normal to an (n − 1)-flat. In three dimensions if a line PN is perpendicular to two intersecting lines NA, NB at their point of intersection N, it is perpendicular to every line through N in the plane determined by NA and NB, and is said to be perpendicular to the plane ANB; and conversely, all lines through N perpendicular to NP lie in one plane which is perpendicular to NP.

We shall consider first all lines, planes, etc., passing through a fixed point O. The number of degrees of freedom of a p-flat which has to pass through O is $p(n - p)$. One condition is required in order that two given lines through O should be perpendicular. Let one line a be fixed, then a line l through O perpendicular to a has $n - 2$ degrees of freedom and may therefore move in a region of $n - 1$ dimensions. Let l_1 and l_2 be any two lines through O perpendicular to a, then all lines of the plane $(l_1 l_2)$ which pass through O are perpendicular to a. In this plane we have then just a single infinity of lines through O perpendicular to a. Let l_3 be a third line through O perpendicular to a and not lying in the plane $(l_1 l_2)$. l_1, l_2, l_3 determine a three-flat; let b be any other line of this three-flat passing through O. The planes $(l_1 l_2)$ and $(l_3 b)$ intersect in a line c which lies in $(l_1 l_2)$ and is therefore $\perp a$. Then since c and l_3 are both $\perp a$, and b lies in the plane $(c l_3)$, $b \perp a$. Hence a is perpendicular to every line through O in the three-flat $(l_1 l_2 l_3)$.

Proceeding in this way we obtain $n - 1$ independent lines $l_1, l_2, \ldots, l_{n-1}$, no three lying in one plane, and all $\perp a$; and every line through O in the $(n - 1)$-flat determined by these lines is $\perp a$. Hence through any point O of a straight line a there is a unique $(n - 1)$-flat which is normal to a; and at any point O in a given $(n - 1)$-flat there is a unique line a which is $\perp$ every line of the $(n - 1)$-flat which passes through O. The normal $(n - 1)$-flat to the line a at O contains every p-flat which is normal to a at O.

2. System of n Mutually Orthogonal Lines. Complete Orthogonality. Through any point O we can find n lines all mutually perpendicular. Starting with a line l_1, all lines through O $\perp$ l_1 lie in the normal $(n - 1)$-flat to l_1 at O. Let l_2 be one of these lines. Then all lines $\perp$ both l_1 and l_2 at O lie in the $(n - 2)$-flat which is the intersection of the normal $(n - 1)$-flats to l_1 and l_2. Let l_3 be one of these lines. Proceeding in this way we shall arrive at a system of n lines $l_1, l_2, \ldots, l_n$, all mutually perpendicular. p of the lines $l_1, l_2, \ldots l_p$ determine a p-flat S_p, and the remaining $n - p$ lines determine an $(n - p)$-flat S_{n-p}. These two flats, which intersect only at O, have the property that every line of S_p through O is perpendicular to every line of S_{n-p} through O. The two flats are said to be *completely orthogonal.*

In two dimensions two lines at right angles, and in three dimensions a plane and a normal line, are completely orthogonal, but two planes cannot be completely orthogonal. In the case of two orthogonal planes in three dimensions only one line in each plane is orthogonal to every line in the other.

In four dimensions the assemblage of all lines through a point O of a plane α perpendicular to α is another plane α' which is completely orthogonal to α. If a is any line of α, and a' any line of α', through O, the plane (aa') is perpendicular to both α and α' in the ordinary three-dimensional sense.

When two lines in space do not intersect we can still say that they are orthogonal when two lines, parallel to them, intersect at right angles. Similarly two planes in S_n $(n > 4)$, which have no point in common, may still be said to be completely orthogonal when two planes, completely parallel to them respectively, and having a common point, are completely orthogonal.

3. Orthogonality in Three Dimensions in Relation to the Absolute; Absolute Poles and Polars. The orthogonality of lines and planes in three dimensions may be explained with reference to their points and lines at infinity. Consider first two lines a, a' and use ordinary rectangular co-ordinates.

Let the two lines pass through the origin and have direction-cosines (l, m, n) and (l', m', n'). Writing x/t, y/t, and z/t instead of x, y, and z, so that x, y, z, t are now the homogeneous cartesian co-ordinates, the equation of the plane at infinity is $t = 0$. The co-ordinates of the points at infinity on the two lines are $(l, m, n, 0)$ and $(l', m', n', 0)$. The condition that the two lines should be orthogonal is $ll' + mm' + nn' = 0$. But this is the condition that the two points $(l, m, n, 0)$ and $(l', m', n', 0)$ should be conjugate with regard to the virtual conic or circle $x^2 + y^2 + z^2 = 0, t = 0$. Similarly two planes in S_3 are orthogonal when their lines at infinity are conjugate with regard to the virtual circle at infinity; and a line and a plane are orthogonal when the point at infinity on the line is the pole of the line at infinity on the plane. This conic, or rather the degenerate quadric surface of which this conic forms the envelope and the plane at infinity taken double forms the locus, is called the *absolute*. In two dimensions it reduces to just two imaginary points called the "circular points at infinity," together with the line at infinity taken twice. In non-euclidean geometry the absolute is a non-degenerate conic, real for hyperbolic geometry and virtual for elliptic geometry.

By viewing the absolute as a degenerate quadric surface, and not simply as a conic in the plane at infinity, we may obtain the absolute polars of finite elements. The polar of any finite point with regard to the absolute, which as a locus consists of just the plane at infinity taken double, is the plane at infinity itself; the polar of a point at infinity P_∞ is any plane passing through the polar line (at infinity) of P_∞ with regard to the circle at infinity. Hence the absolute pole of an ordinary plane α is the point at infinity which is the pole of the line at infinity in α with regard to the circle at infinity. The absolute polar of a line is the line at infinity polar with regard to the circle at infinity of the point at infinity on the line. The absolute pole of an ordinary plane is the point at infinity on any of its normals, and the absolute polar of an ordinary line is the line at infinity on any of its normal planes. When we speak of the absolute polar of a finite element we shall mean the polar with regard to the degenerate quadric, but when we speak of the polar of an element at infinity we shall mean the polar with regard to the virtual quadric at infinity.

We may then state the relations as follows. A line is perpendicular to a plane when it passes through the absolute polar of the plane, and reciprocally the plane contains the absolute polar of the line. Two lines are orthogonal when each cuts the absolute polar of the other. Two planes in S_3 are orthogonal when each contains the absolute pole of the other.

4. **The Absolute in Four Dimensions.** In euclidean space of four dimensions the absolute is a degenerate quadric consisting as a locus of the hyperplane at infinity taken twice and as an envelope of a virtual quadric in the hyperplane at infinity. In S_4 two lines are orthogonal when their points at infinity are conjugate with regard to the absolute; a line and a hyperplane, when the point at infinity on the line is the pole of the plane at infinity on the hyperplane; two hyperplanes when their planes at infinity are conjugate, i.e. when the normal to one hyperplane lies in the other.

Two arbitrary planes α, β in S_4 cut the hyperplane at infinity in two arbitrary lines α_∞, β_∞. In general these lines do not intersect, and the two planes have one finite point in common and no common points at infinity. If the two lines at infinity intersect, either this common point at infinity is the only common point of the two planes, and they are half-parallel; or they have also a finite common point and therefore a line in common and are then situated in the same three-flat. If the two lines at infinity coincide the two planes are completely parallel.

5. If the two lines at infinity are mutual polars, the polar planes of all points on one line pass through the other, and every point on the one line is conjugate to every point on the other. In this case the two planes are completely orthogonal; all the normals to the one plane at their common point lie in the other plane, and every line in the one plane is orthogonal to every line in the other plane.

In general the polar plane of a point on the one line cuts the other line, so that to every point on the one line there corresponds just one point on the other which is conjugate to it. Thus in the case of two arbitrary planes through O, to every line through O in the one plane there corresponds just one line through O in the other plane such that the two lines are orthogonal.

If, however, there should happen to exist a point a_∞ in α_∞ such that its polar plane passes through β_∞, then every point in β_∞ is conjugate to a_∞, and there corresponds in α a line a which is orthogonal to every line in β. Let α'_∞ and β'_∞ be the polars of α_∞ and β_∞. Then since α'_∞ lies in the polar plane of every point of α_∞, and therefore in the polar plane of a_∞, which contains β_∞, the lines α'_∞ and β_∞ lie in the same plane and intersect in a point b_∞; then the polar plane of b_∞ passes through α_∞, and β'_∞ cuts α_∞ in a_∞. Hence the two planes α, β are reciprocally related: in α there is a line a which is orthogonal to every line in β, and in β there is a line b which is orthogonal to every line in α. The two planes are then said to be *half-orthogonal.* The condition for this therefore is that the line at infinity on each should intersect the polar of the line at infinity on the other, or that each plane should be half-parallel to the plane which is completely orthogonal to the other. Two planes which are orthogonal in the ordinary sense in space of three dimensions evidently satisfy this condition and are half-orthogonal, but we may have half-orthogonal planes in space of higher dimensions which meet in just one point, or which have no point in common.

The lines at infinity α'_∞ and β'_∞ are the absolute polars of α and β. We may then state the conditions for orthogonality of two planes in the form: two planes in S_4 are completely orthogonal when each contains the absolute polar of the other; they are half-orthogonal when each cuts the absolute polar of the other.

Every plane through a normal line to a given plane is half-orthogonal to the given plane; hence if α and β are completely orthogonal at O, so that α'_∞ coincides with β_∞, and β'_∞ with α_∞, every plane through O which cuts α in a straight line is half-orthogonal to β, and every plane which cuts both α and β in straight lines is orthogonal to both α and β in the ordinary three-dimensional sense. If α and β are half-parallel, so that α_∞ cuts β_∞ in a point C_∞, the polar plane of C_∞ passes through both α'_∞ and β'_∞ and therefore these two lines also intersect, in a point C'_∞, say; hence if α' and β' are the planes through O which are completely orthogonal to α and β, α' and β' are also half-parallel. If the four lines α_∞, β_∞, α'_∞, β'_∞ intersect in this order, forming a skew quadrilateral (Fig. 7), the two planes α and β are both half-parallel and half-orthogonal provided they have no finite point in common; if they have a finite point in common they intersect in a line and are orthogonal in the ordinary sense in three dimensions. (If α is a real plane it is not possible for α_∞ to intersect α'_∞ for then we should have a line in α orthogonal to itself.)

6. Orthogonality in S_n. We can now extend these ideas to space of n dimensions. Consider a p-flat α and a q-flat β ($q \leqq p$) having not more than a point in common, so that $n \geqq p + q$. Their regions at infinity are a $(p - 1)$-flat α_∞ and a $(q - 1)$-flat β_∞ while the absolute is a virtual hyperquadric in the $(n - 1)$-flat at infinity. The absolute polar of any point P_∞ in α_∞ is an $(n - 2)$-flat, and if p independent points are chosen on α_∞ their absolute polars intersect in an $(n - p - 1)$-flat α'_∞, the absolute polar of α_∞. Similarly the absolute polar of β_∞ is an $(n - q - 1)$-flat β'_∞. If α is real, α_∞ and α'_∞ can have no common point. Also α'_∞ has in general no region in common with β_∞ (since $n - p - 1 + q - 1 < n - 1$), but β'_∞ has in common with α_∞ a $(p - q - 1)$-flat at least.

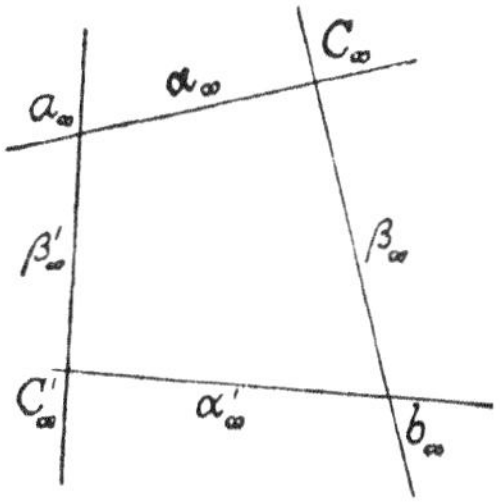

FIG. 7.

If β'_∞ contains α_∞, then α'_∞ contains β_∞; α and β are then *completely orthogonal*, and every flat in α is completely orthogonal to every flat in β.

In the general case there is in α at least a $(p - q)$-flat which is completely orthogonal to β, and every flat which lies in this $(p - q)$-flat or is completely parallel to it is completely orthogonal to β. The absolute polar $(n - 2)$-flat of any point P_∞ in α_∞ cuts β_∞ in a $(q - 2)$-flat. Hence to every line in α there corresponds a series of parallel $(q - 1)$-flats in β normal to it, to every r-flat in α ($r < q$) there corresponds a series of parallel $(q - r)$-flats in β completely orthogonal to it. Similarly to every r-flat in β there corresponds a series of parallel $(p - r)$-flats in α completely orthogonal to it.

7. Degrees of Orthogonality. If α'_∞ and β_∞ have a common r-flat b_∞, the polar $(n-2)$-flat of every point on α_∞ contains b_∞. β'_∞, the polar $(n-q-1)$-flat of β_∞, is determined by the intersection of the polar $(n-2)$-flats of any q independent points on β_∞; $r+1$ of these can be taken in b_∞, and the polar $(n-2)$-flats of these all contain α_∞; the polar $(n-2)$-flats of the remaining $q-r-1$ points intersect in an $(n-q+r)$-flat which cuts α_∞ in a $(p-q+r)$-flat a_∞, i.e. β'_∞ cuts α_∞ in a $(p-q+r)$-flat a_∞. Every point in a_∞ is conjugate to every point in β_∞, and every point in b_∞ is conjugate to every point in α_∞. Hence in this case α contains a series of parallel $(p-q+r+1)$-flats which are completely orthogonal to β, and β contains a series of parallel $(r+1)$-flats completely orthogonal to α. When $r+1=q$, α and β are completely orthogonal. When $r+1=0$ they are in general position; there is then no line in β orthogonal to every line of α, although there is always a $(p-q)$-flat in α which is completely orthogonal to β if $p > q$. Thus in three dimensions, with a line and a plane in general position, there is just one direction in the plane which is orthogonal to the given line. The *degree of orthogonality* of a p-flat α and a q-flat β $(p \geqq q)$ may be represented by the fraction $(r+1)/q$, where r is the dimension of the region common to α'_∞ and β_∞, i.e. the denominator represents the dimensions of the lower-dimensioned space β, and the numerator is the highest dimension of a flat lying in β and completely orthogonal to α.

8. Consider now the case in which α and β have an m-flat μ in common. Their regions at infinity α_∞ and β_∞ then intersect in an $(m-1)$-flat μ_∞. The absolute polars α'_∞ and β'_∞ have in common μ'_∞, the absolute polar $(n-m)$-flat of μ_∞. If α'_∞ cuts β_∞ in an r-flat, this r-flat cuts μ_∞ in an $(m+r-q)$-flat, but since μ_∞ and μ'_∞ can have no common point, α'_∞ cannot cut μ_∞, hence $r < q-m$. The highest degree of orthogonality is therefore $(q-m)/q$, which occurs when $r=q-m-1$. Then α'_∞ and β_∞ have a common $(q-m-1)$-flat, and α_∞ and β'_∞ have a common $(p-m-1)$-flat. In this case α contains a series of parallel $(p-m)$-flats completely orthogonal to β, and β contains a series of parallel $(q-m)$-flats completely orthogonal to α.

Conversely, in order that a p-flat and a q-flat ($q \gtrless p$) should have orthogonality of degree r/q they must not have in common a space of higher dimensions than $q - r$; the space which contains them both must be of dimensions not lower than $p + r$. For complete orthogonality they must not have more than a point in common, and must not be contained in a space of dimensions lower than $p + q$.

In three dimensions a line and a plane may be completely orthogonal, but two planes can only be half-orthogonal.

9. A line perpendicular to an S_{n-1} in S_n passes through the absolute pole of S_{n-1}. Hence all lines in S_n which are orthogonal to the same S_{n-1} are parallel. Similarly an S_p which is completely orthogonal to an S_q in S_{p+q} passes through the absolute polar $(p - 1)$-flat at infinity of S_q. Hence *all p-flats which are completely orthogonal to the same q-flat and lie in a given $(p + q)$-flat are completely parallel.*

10. If two linear spaces are completely orthogonal, every linear space contained in one is completely orthogonal to every linear space contained in the other.

Consider a p-flat α and a q-flat β ($p \gtrless q$) orthogonal of degree $(r + 1)/q$, and having not more than a point in common, so that α'_∞ and β_∞ have a common r-flat b_∞, and α_∞ and β'_∞ have a common $(p - q + r)$-flat a_∞.

Let γ be any s-flat lying in β. Its region at infinity γ_∞ lies in β_∞ and cuts b_∞ and therefore α'_∞ in a k-flat. (i) Let $s > r$ and also $s \gtrless q - r$. Then $s + r - q \gtrless k \gtrless r$. Hence γ is orthogonal to α of degree between $(s + r - q + 1)/s$ and $(r + 1)/s$ inclusive. (ii) Let $q - r > s > r$, so that $r < \frac{1}{2}q$. The lower limit of orthogonality is then zero. (iii) Let $r \gtrless s \gtrless q - r$, so that $r \gtreqless \frac{1}{2}q$. Then $s + r - q \gtrless k \gtrless s$, and the limits of orthogonality are $(s + r - q + 1)/s$ and 1. (iv) Let $s \gtrless r$ and also $s < q - r$. Then all degrees of orthogonality are possible.

Next consider an s-flat γ lying in α, and (i) let $s \gtrless q$. The $(n - s - 1)$-flat γ'_∞ contains α'_∞ and therefore has in common with β_∞ an r-flat at least. Hence γ is orthogonal to β of degree $(r + 1)/q$ at least, and may be completely orthogonal to β. (ii) Let $q - r < s < q$ and also

$$s - 1 > p - q + r.$$

γ_∞ lies in α_∞ and cuts a_∞ and therefore β'_∞ in a k-flat where $s - q + r \gtrless k \gtrless p - q + r$. Then γ is orthogonal to β of degree between $(s - q + r + 1)/s$ and $(p - q + r + 1)/s$. (iii) If $q - r < s < q$ and $s - 1 < p - q + r$, the upper limit of orthogonality becomes 1. (iv) If $s \gtrless q - r$ and $s - 1 > p - q + r$ the lower limit of orthogonality becomes zero, and (v) if $s \gtrless q - r$ and $s - 1 < p - q + r$ all degrees of orthogonality are possible.

Other cases in which α and β intersect can be worked out in a similar way.

As a further example, suppose α and β are completely orthogonal, so that α'_∞ contains β_∞, and β'_∞ contains α_∞; and consider an s-flat γ which contains β. γ_∞ contains β_∞ and therefore cuts α'_∞ in a $(q - 1)$-flat at least. Hence if $s \gtrless p$, γ is orthogonal to α of degree at least q/s. It is completely orthogonal only if γ_∞ lies in α'_∞; this requires that $n - p - 1 \gtrsim s - 1$, i.e. $n \gtrsim s + p$, as of course γ and α must have no more than a point in common. If $s > p$, we note that γ'_∞ lies in β'_∞ and cuts α_∞ in at least a $(p + q - s - 1)$-flat, and the orthogonality is of degree $(p + q - s)/p$ at least. Consider now γ containing α. γ_∞ contains α_∞ and therefore cuts β'_∞ in a $(p - 1)$-flat at least. Then γ is orthogonal to β of degree p/s at least.

REFERENCE

SOMMERVILLE, D. M. Y. The elements of non-euclidean geometry. London: Bell, 1914.

CHAPTER IV

DISTANCES AND ANGLES BETWEEN FLAT SPACES

1. Two linear spaces which have their highest degree of intersection determine an angle, and this angle determines completely the shape of the figure consisting of the two spaces. For example, two straight lines in a plane determine a plane angle; two planes in 3-space determine a dihedral angle, which can be measured by means of a plane angle. If the figure is moved about rigidly this angle remains fixed; or, to put it more geometrically, if the figure is referred to a system of rectangular co-ordinates and we transform to any other system of rectangular co-ordinates there is an analytical expression corresponding to the angle which remains *invariant.*

2. **Angle between two Completely Intersecting Spaces.** We have then to show first that two flat spaces S_p and S_q ($p \geq q$) which intersect in an S_{q-1} and therefore lie in the same S_{p+1} have a single mutual invariant.

If S_{q-1} is not simply a point there is a unique S_{p-q+2} through any point O of S_{q-1} which is completely orthogonal to S_{q-1}. This cuts S_q in a line and S_p in an S_{p-q+1}. Conversely, if the line and the S_{p-q+1} are given, intersecting in O, they determine a unique S_{p-q+2} containing them both; at O, and in an arbitrary S_{p+1} containing S_{p-q+2}, there is a unique S_{q-1} completely orthogonal to S_{p-q+2}, and a unique S_p containing S_{q-1} and S_{p-q+1}. We have therefore reduced the problem to determining the angle between a line S_1 and an S_r which cut in a point. In S_r there is a unique S_{r-1} orthogonal to S_1, and a unique line l orthogonal to S_{r-1}. The two lines S_1 and l then determine the unique angle.

If now a similar construction is made with another point O′ on S_{q-1}, then we have at O a line l and a ($p - q + 1$)-flat S_{p-q+1}, both completely orthogonal to S_{q-1}, and also at O′ a line l' and an S'_{p-q+1}, both completely orthogonal to S_{q-1}. But l and l' both lie in S_q and are therefore parallel, also S_{p-q+1} and S'_{p-q+1} both lie in S_p and are therefore completely parallel. Hence the angles determined at O and O′ are equal.

3. **Distance between Non-Intersecting Spaces.** If two p-flats α and β are completely parallel we have to regard their angle as being zero. In the $(p+1)$-flat which contains them there passes through any point O a single line which is normal to both α and β. If two normals are drawn through different points O, O′, cutting α and β in A, B and A′, B′, then ABB′A′ is a rectangle and AB = A′B′. This distance AB is called the distance between the parallel p-flats. We can similarly, with a slight modification, define the distance between a p-flat and a q-flat which are completely parallel.

4. If S_p and S_q have no common point, they are contained in the same S_{p+q+1}. Let $S_{(p-1)\infty}$ and $S_{(q-1)\infty}$ be their infinite regions; these lie in the region at infinity $S_{(p+q)\infty}$ and have no common point. Then S_p and $S_{(q-1)\infty}$ determine a $(p+q)$-flat A_{p+q} which contains S_p

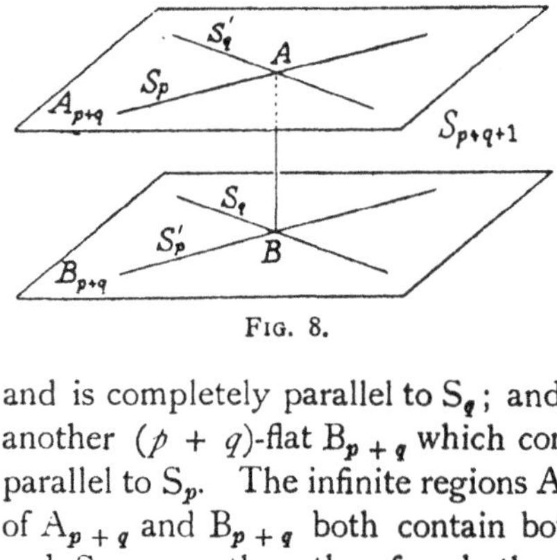

FIG. 8.

and is completely parallel to S_q; and S_q and $S_{(p-1)\infty}$ determine another $(p+q)$-flat B_{p+q} which contains S_q and is completely parallel to S_p. The infinite regions $A_{(p+q-1)\infty}$ and $B_{(p+q-1)\infty}$ of A_{p+q} and B_{p+q} both contain both of the regions $S_{(p-1)\infty}$ and $S_{(q-1)\infty}$; they therefore both coincide with $S_{(p+q)\infty}$, and therefore A_{p+q} and B_{p+q} are completely parallel. The distance between A_{p+q} and B_{p+q} is then defined to be the distance between S_p and S_q.

S_p determines with $A'_{0\infty}$, the absolute pole of $A_{(p+q-1)\infty}$, a $(p+1)$-flat which cuts B_{p+q} in a p-flat S'_p and S_q in a point B; and similarly S_q determines with $B'_{0\infty}$, the absolute pole of $B_{(p+q-1)\infty}$, a $(q+1)$-flat which cuts A_{p+q} in a q-flat S'_q and S_p in a point A. AB is normal to both A_{p+q} and B_{p+q} and therefore also to S_p and S_q, and the construction shows that it is in general unique. In general therefore *two flats which have no point in common have a unique common normal.*

5. If S_p and S_q are parallel of degree $(r+1)/q$ so that they have no finite point in common and their infinite regions intersect in an r-flat $S_{r\infty}$, they are contained in the same $(p+q-r)$-flat. Then S_p and $S_{(q-1)\infty}$ determine a $(p+q-1-r)$-flat $A_{p+q-r-1}$ which contains S_p and is completely parallel to S_q; and S_q and $S_{(p-1)\infty}$ determine another $(p+q-r-1)$-flat $B_{p+q-r-1}$ which contains S_q and is completely parallel to S_p; also $A_{p+q-r-1}$ is completely parallel to $B_{p+q-r-1}$. S_p determines with $A'_{0\infty}$, the absolute pole of $A_{(p+q-r-2)\infty}$, a $(p+1)$-flat which cuts $B_{p+q-r-1}$ in a p-flat S'_p and S_q in an $(r+1)$-flat B_{r+1}; and similarly S_q determines with $B'_{0\infty}$, the absolute pole of $B_{(p+q-r-2)\infty}$, a $(q+1)$-flat which cuts $A_{p+q-r-1}$ in a q-flat S'_q and S_p in an $(r+1)$-flat A_{r+1}. Since all the flats contain $S_{r\infty}$, it follows that $B_{r\infty}$ and $A_{r\infty}$, the regions at infinity on B_{r+1} and A_{r+1}, both coincide with $S_{r\infty}$. Therefore A_{r+1} and B_{r+1} are completely parallel. Hence *when two flats are parallel, having a common r-flat at infinity, there is a unique $(r+1)$-flat on each which is completely parallel to the other and the two flats have ∞^{r+1} common normals.*

Two lines in a plane have a single angle, which becomes zero when the lines are parallel; two parallel lines have a single distance. Two lines in general have a single distance, which becomes zero when the lines intersect; and also a single angle, which is equal to the angle between two intersecting lines parallel to the given lines. Thus two lines in general have two mutual invariants. Two planes in three dimensions have a single angle, which is zero when the planes are completely parallel; two completely parallel planes have a single distance.

6. Angles between Two Planes in S_4. Consider now the case of two planes α, β in general. If they have no point in common we have seen that they have a single distance, the distance between two 3-flats, one through each plane and completely parallel to the other. If they are contained in the same 3-flat they have a single dihedral angle. We have now to consider the case where they have just one point O in common. Let a_∞ and b_∞ be their lines at infinity. These do not in general intersect. If they intersect, α and β intersect in a line; if they coincide, α and β are completely parallel. The absolute polars of a_∞ and b_∞ are two lines a'_∞ and b'_∞. If α and β are real, a_∞ and a'_∞ cannot intersect. If a_∞ cuts b'_∞, reciprocally a'_∞ cuts b_∞, and the two planes are half-orthogonal. If a_∞ coincides with b'_∞, and reciprocally a'_∞ coincides with b_∞, the two planes are completely orthogonal. In the general case a_∞, b_∞, a'_∞, b'_∞ are four skew lines. Take any other plane γ through O, and let c_∞ be its line at infinity and c'_∞ the absolute polar of this line. If c_∞ cuts a_∞, γ cuts α in a line; if c_∞ cuts a'_∞, γ and α are half-orthogonal. Hence if c_∞ cuts both a_∞ and a'_∞, γ and α are orthogonal in three dimensions. Through any point B_∞ on b_∞ there passes just one line which cuts both a_∞ and a'_∞, viz. the line of intersection of the planes $B_\infty a_\infty$ and $B_\infty a'_\infty$. Hence through any line m of β, through O, there passes just one plane which cuts α orthogonally in a straight line, l say; this plane is not in general orthogonal to β. If, however, it is possible to find a common transversal $A_\infty B_\infty A'_\infty B'_\infty$ of all four lines a_∞, b_∞, a'_∞, b'_∞ we obtain a plane through O which cuts both α and β orthogonally in straight lines l and m. Now this is in general possible in two ways. The assemblage of all transversals of three skew lines a_∞, b_∞, a'_∞ is a *regulus*, or system of lines on a ruled surface of the second degree (hyperboloid or paraboloid). On this surface there is a second set of lines all mutually skew, and every line of the one system cuts every line of the other system. The fourth line b'_∞ cuts the surface in two points B'_1 and B'_2, and through each of these there passes a line of the second system and these therefore cut the first three lines in points A_1, B_1, A'_1 and A_2, B_2 A'_2 (Fig. 9). Hence we have two lines d_1 and d_2 cutting all four lines.

We have therefore through O two planes, δ_1 and δ_2 say, each cutting α and β orthogonally in straight lines, say l_1 and l_2 on α, and m_1 and m_2 on β. The two planes therefore determine two angles, $\phi_1 = \measuredangle (l_1 m_1)$ and $\phi_2 = \measuredangle (l_2 m_2)$, the angles which they cut out on the two common orthogonal planes δ_1 and δ_2.

7. If we transform the figure by taking the absolute polar of every point and line, a_∞ and a'_∞, b_∞ and b'_∞ are interchanged; d_1 and d_2 are changed into their absolute polars d'_1 and d'_2. But since properties of incidence are invariant for this transformation, d'_1 and d'_2 must also cut the four lines, and since there are only two common transversals, d'_1 must coincide with d_2 and d'_2 with d_1. Hence d_1 and d_2 are absolute polars, and the corresponding planes δ_1 and δ_2 are completely orthogonal.

8. The existence of these two angles can be shown also as follows. If OP is any line in α and OQ any line in β, both through O, the angle POQ can never be zero and must therefore have a certain minimum value. If A_1OB_1 (Fig. 10) represents this minimum angle, the plane $A_1OB_1 (\equiv \delta_1)$ must cut both

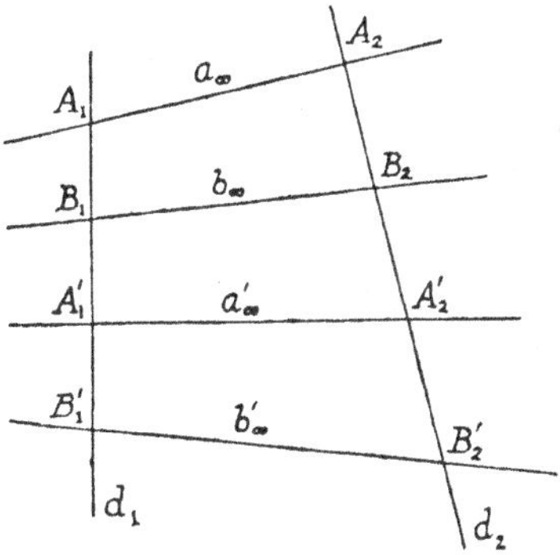

FIG. 9.

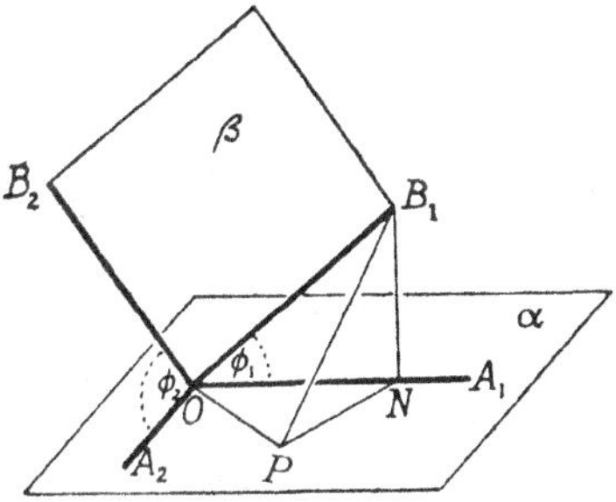

FIG. 10.

α and β orthogonally, for consider another plane through OB_1 cutting α in OP, then when the plane A_1OB_1 is orthogonal to α the angle $B_1OP >$ angle B_1OA_1. Having shown the existence of one plane cutting both α and β orthogonally and forming a minimum angle A_1OB_1, a second plane can be found having the same property. In α take the line $OA_2 \perp OA_1$, and in β the line $OB_2 \perp OB_1$. Then since $\alpha \perp \delta_1$, $OA_2 \perp \delta_1$, and therefore $OA_2 \perp OB_1$. Similarly $OB_2 \perp OA_1$. The two lines OA_1, OB_1 are both perpendicular to the two lines OA_2, OB_2, and the plane $A_2OB_2 (\equiv \delta_2)$ is normal to OA_1 and OB_1 and therefore cuts both α and β orthogonally. The acute angle $A_2OB_2 (\equiv \phi_2)$ is therefore also a minimum. Also the planes δ_1 and δ_2 are completely orthogonal.

If OP_1 is any line of α there is one plane through OP_1 which cuts β orthogonally in OQ_1, say. Then there is one plane through OQ_1 which cuts α orthogonally in OP_2, say, and so on. Now since the plane $OQ_1P_2 \perp OP_1P_2$, the angle $Q_1OP_2 <$ the angle P_1OQ_1. Hence we have a sequence of lines OP_1, OP_2, . . . on α and a sequence of lines OQ_1, OQ_2, . . . on β such that

$$\angle P_1OQ_1 > Q_1OP_2 > P_2OQ_2 > \ldots$$

The sequence P_1OQ_1, P_2OQ_2, . . . therefore tends to a finite limit, which is A_1OB_1. Similarly the sequence in the ascending order tends to a finite limit which is A_2OB_2. Thus while each of the two angles ϕ_1 and ϕ_2 is a minimum value of the angle POQ, where OP and OQ are any lines in α and β respectively, ϕ_1 is a minimum and ϕ_2 a maximum in the series of angles P_1OQ_1 for which the plane P_1OQ_1 is always $\perp \alpha$.

9. We see also that if $\phi_1 = \phi_2$, all the intermediate angles P_1OQ_1, etc., are equal, and each of the planes P_1OQ_1 cuts both α and β orthogonally. In this case the two planes are said to be *isocline*. The lines at infinity a_∞ and b_∞, and their absolute polars a'_∞ and b'_∞ all belong to the same regulus, and are cut by all the generators of the other system.

The theory of isocline planes through a point is the same as the theory of equidistant straight lines in elliptic geometry of three dimensions, usually called Clifford's parallels or paratactic lines.*

Thus it may be proved that if the planes α and β are isocline, so also are any two planes which cut them both orthogonally. Through any line OA which cuts a given plane α in O there pass two planes isocline to α. Let the plane through $OA \perp \alpha$ cut α in ON, and let $\angle NOA = \theta$. Denote the plane NOA by δ, and let δ' be the plane through O completely orthogonal to δ. δ' cuts α in the line $ON' \perp ON$. In δ' construct the two lines OB_1 and OB_2, each making with ON' the angle θ. Then the planes AOB_1 and AOB_2 are each isocline to α.

10. Returning now to the general case, two planes α and β through a point O have two minimum angles $A_1OB_1 = \phi_1$ and

* See Sommerville, "The Elements of Non-euclidean Geometry," chap. iii.

$A_2OB_2 = \phi_2$. If $\phi_1 = 0$ the lines OA_1 and OB_1 coincide and the planes have the line OA_1 in common; the other angle is the ordinary dihedral angle between the two planes. If also $\phi_2 = 0$ the planes coincide. If $\phi_2 = 90°$ the planes are half-orthogonal, and if also $\phi_1 = 90°$ they are completely orthogonal.

11. If the two planes α, β have no point in common and have only a common containing 5-flat, they have a minimum distance d, and also two minimum angles ϕ_1 and ϕ_2 which may be defined as being equal to those belonging to two planes through a point O which are completely parallel to the given planes. If $\phi_1 = 0$, α and β are half-parallel; if also $\phi_2 = 0$ they are completely parallel; and they coincide only if d, ϕ_1 and ϕ_2 all vanish. Thus two planes in S_5 have three mutual invariants, one distance and two angles.

12. Angles between two Linear Spaces. Two 3-flats which lie in S_6, and therefore have a point in common, cut the hyperplane at infinity in two planes having no point in common. These two planes have three mutual invariants, all distances, and these measure three angles which form the mutual invariants of the two 3-flats.

It can be proved by induction that two p-flats in S_{2p}, and therefore having a point in common, have p mutual invariants, all angles. For assuming this for p, two p-flats in S_{2p+1}, having no point in common, have $p + 1$ mutual invariants, p angles and one distance. Then two $(p + 1)$-flats in S_{2p+2}, having just one point in common, cut the hyperplane at infinity in two p-flats which have no point in common and have therefore $p + 1$ mutual invariants; these determine the $p + 1$ mutual invariants of the two $(p + 1)$-flats.

13. Consider now an S_p and an S_q $(p > q)$ having just one point in common, and lying in an S_{p+q}. From every point of S_q there can be drawn one and only one line cutting S_p at right angles. The assemblage of the feet of the perpendiculars is an S'_q which is the projection of the given S_q on the S_p. The assemblage of all the perpendiculars forms an S_{2q} containing S_q and S'_q. The assemblage of all the perpendiculars to S_{2q} in the S_{p+q} at points of S'_q is the S_p. Thus the S_p is determined when the S_q and S'_q are given, and the mutual invariants of the S_p and S_q are the same as those of the S_q and S'_q, q in number.

14. Consider lastly an S_p and an S_q having in common an S_r and all lying in an S_{p+q-r} ($p > q > r$). As before this may be reduced to the S_q and its projection S'_q on S_p, so that we have only to consider two S_q's having in common an S_r and lying in an S_{2q-r}. At any point of S_r there is a unique S_{2q-2r} in the S_{2q-r} absolutely orthogonal to S_r, and this cuts the S_q's in S_{q-r}'s. We have then two S_{q-r}'s in S_{2q-2r} and having one point in common, and these determine the original S_p and S_q. Hence we have in this case $q - r$ mutual invariant angles. An S_p and an S_q ($p \geqq q$) both lying in an S_n ($n \lesseqgtr p+q$) have $n - p$ mutual invariant angles; if $n \geqq p + q + 1$, they have q angles and one distance.

15. The actual construction of the q angles between an S_p and an S_q($p \geqq q$) which have one point O in common can be obtained as follows. If OP is any line in S_p and OQ any line in S_q the angle POQ has a certain minimum value since it cannot be zero, S_p and S_q having no common line. Let $P_1OQ_1 = \theta_1$ be this minimum angle. Then the plane P_1OQ_1 is half-orthogonal to both S_p and S_q, and θ_1 is one of the angles between these two flats.

Now at O there is in S_p a unique S_{p-1} normal to OP_1, and in S_q a unique S_{q-1} normal to OQ_1, and if OP is any line in S_{p-1} and OQ any line in S_{q-1} the angle POQ has again a minimum value, say $P_2OQ_2 = \theta_2$, and the plane P_2OQ_2 is half-orthogonal to both S_{p-1} and S_{q-1}. Since P_1OQ_1 is half-orthogonal to S_p which contains the plane P_1OP_2, it is half-orthogonal to P_1OP_2; and since $OP_2 \perp OP_1$ it follows that $OP_2 \perp P_1OQ_1$. Similarly $OQ_2 \perp P_1OQ_1$. Hence P_2OQ_2 and P_1OQ_1 are completely orthogonal. In the plane P_2OQ_2 let $OP' \perp OP_2$ and $OQ' \perp OQ_2$. Then $OP' \perp S_{p-1}$ and also $\perp OP_1$, hence $OP' \perp S_p$, and similarly $OQ' \perp S_q$. Therefore P_2OQ_2 is half-orthogonal to both S_p and S_q, and the angle θ_2 is a second angle between these two flats.

Take next in S_p the S_{p-2} which is absolutely orthogonal to the plane P_1OP_2, and in S_q the S_{q-2} orthogonal to Q_1OQ_2. Then we obtain again a minimum angle between lines of S_{p-2} and S_{q-2}, viz. $P_3OQ_3 = \theta_3$, and it is proved as before that

P_3OQ_3 is half-orthogonal to S_{p-2} and S_{q-2} and hence to S_p and S_q. Also the three planes P_1OQ_1, P_2OQ_2, P_3OQ_3 are mutually completely orthogonal.

This process can be continued until we have obtained $q - 1$ angles. The next step gives an S_{p-q+1} orthogonal to the $(q - 1)$-flat $OP_1P_2 \ldots P_{q-1}$ and a line l normal to the $(q - 1)$-flat $OQ_1Q_2 \ldots Q_{q-1}$. There is then one line in S_{p-q+1} which makes a minimum angle with the line l, and we then obtain q angles.

A clearer picture of this is obtained when we consider projections (see § 18).

If S_p and S_q have an S_r common and are therefore contained in an S_{p+q-r}, at any point O of S_r there is a unique S_{p+q-2r} completely orthogonal to S_r. This cuts S_p in an S_{p-r} and S_q in an S_{q-r}, and these two spaces have $q - r$ angles which are the angles between S_p and S_q. The other r angles which should exist are now zero.

16. If S_p and S_q $(p \geqq q)$ have just one common point O, and S_{n-p} and S_{n-q} are the spaces through O completely orthogonal to them, these two spaces intersect in an S_{n-p-q} and have therefore $(n - p) - (n - p - q) = q$ angles; these are respectively equal to the q angles between S_p and S_q. If S_p and S_q have an S_r in common and still lie in S_n, so that they have $q - r$ angles, the normal spaces S_{n-p} and S_{n-q} both lie in the S_{n-r} which is normal to S_r and have therefore a common $S_{n-p-q+r}$; they have therefore also $(n - p) - (n - p - q + r) = q - r$ angles, which are equal to the corresponding angles between S_p and S_q.

17. If S_p and S_q $(p \geqq q)$ have just one point O in common and lie in an S_n, so that $n \geqq p + q$, and S_{n-p} is the $(n - p)$-flat normal to S_p at O, $n - p \geqq q$, therefore S_{n-p} and S_q have q angles; these are the complements of the angles between S_p and S_q. Similarly if S_p and S_q have an S_r in common and lie in an S_{p+q-r}. Then the S_{n-p}, i.e. S_{q-r}, normal to S_p, makes $q - r$ angles with S_q which are the complements of the angles between S_p and S_q. But if $p + q > n > p + q - r$, so that $q > n - p > q - r$, the normal S_{n-p}, which cannot meet S_q in more than a point, has $n - p$ angles with S_q, exceeding therefore the number $q - r$. In this case the S_{n-r} normal to

S_r contains S_{n-p} and cuts S_q in an S_{q-r}. The $q - r$ angles between S_q and this S_{q-r} are the complements of those between S_p and S_q.

18. Projections. In the general conception of projection there is a fixed point O, the *centre of projection*, and a fixed $(n - 1)$-flat π, the *hyperplane of projection*, not passing through O, and the projection of a point P on to π is the point P′ in which the line OP cuts π. The projection of a line l is the line of intersection of the plane Ol with π, and in general the projection of a p-flat α is the p-flat α' in which the $(p + 1)$-flat Oα cuts π. The projection of a p-flat α may be a $(p - 1)$-flat, in the case where α passes through O; α' is then just the intersection of α with π.

In this conception of projection every element has a unique projection, but any p-flat α' in π is the projection of any p-flat which lies in the $(p + 1)$-flat Oα'. A more specialised and fruitful conception of projection, which, however, is not now being considered, confines the elements, which are being projected, to an $(n - 1)$-flat not passing through O: this is the basis of projective geometry.

To project on to an $(n - 2)$-flat, as hyperplane of projection π, the centre of projection is replaced by a straight line l, the *axis* of projection, which does not cut π. The projection of a point P is then the point P′ in which the plane lP cuts π. The projection of a p-flat α is the p-flat α' in which the $(p + 1)$-flat $l\alpha$ cuts π.

Similarly to project on to an s-flat π, we take as axis of projection an $(n - s - 1)$-flat S, which does not cut π, and the projection of a point P is the point P′ in which the $(n - s)$-flat SP cuts π. The projection of a p-flat α $(p < s)$ is the p-flat α' in which the $(n - s + p)$-flat Sα cuts π. The projection of a p-flat will be a $(p - r)$-flat if the p-flat cuts S in an $(r - 1)$-flat; a point if it cuts S in a $(p - 1)$-flat.

19. Parallel Projection. In parallel projection the axis of projection is an $(n - s - 1)$-flat at infinity S_∞, the hyperplane of projection being an s-flat π, whose region at infinity does not cut S_∞. The $(n - s)$-flats which project points are all completely parallel to one another. Since S_∞ lies in the hyperplane at infinity, the projection of any element at infinity is itself an element at infinity. If a p-flat and a q-flat ($s > p \leqq q$) are parallel of degree $(r+1)/q$, so that they intersect in an r-flat at infinity, and have no finite point in common, their projections are also parallel of degree $(r+1)/q$. In particular the projection of a parallelotope is a parallelotope.

20. Orthogonal Projection. In orthogonal projection, which is a particular case of parallel projection, the axis of projection is the absolute polar π'_∞ of the s-flat of projection π. The $(n-s)$-flats which project points on to π are completely orthogonal to π and are all completely parallel to one another. The projection of a p-flat α ($p < s$) is always a p-flat, provided α has no degree of orthogonality with π. As in parallel projection in general, the projection of a parallelotope is always a parallelotope.

If S'_q is the projection of S_q on S_p the angles between S_p and S_q are the same as those between S_q and S'_q. In each of these flats we have, by the construction of § 15, a set of q rectangular axes $OP_1, OP_2, \ldots, OP_q$ and $OQ_1, OQ_2, \ldots, OQ_q$, and these are so related that $OP_r \perp OQ_s$ if $r \neq s$, but the angle between corresponding axes OP_r and $OQ_r = \theta_r$.

21. Projection of an Orthotope into an Orthotope. A rectangular parallelotope, i.e. one whose concurrent edges are all mutually at right angles, will be called an *orthotope*. Consider an orthotope of p dimensions ($p < s$) with edges $a_1, a_2, \ldots, a_p$ concurrent at O, and suppose first that its p-flat α does not cut π, or only cuts π in one point. Let O′ be the projection of O, and let α' be the p-flat through O′ completely parallel to α. Then in α' there are p mutually perpendicular axes $O'P'_1, O'P'_2, \ldots, O'P'_p$, and in π thereare also p mutually perpendicular axes $O'Q_1, O'Q_2, \ldots, O'Q_p$, the projections of the former, such that the angles P'_iOQ_j ($i \neq j$) are all right angles, while $P'_iOQ_i = \theta_i$. The angles $\theta_1, \ldots, \theta_p$ are then the p angles between α' and π or between α and π. If then the edges of the orthotope in α are parallel to the corresponding axes in α', the orthotope is projected into an orthotope with edges $a'_1 = a_1 \cos\theta_1$, $a'_2 = a_2 \cos\theta_2, \ldots$, $a'_p = a_p \cos\theta_p$.

If α cuts π in an r-flat, r of the angles $\theta_{p-r+1}, \ldots, \theta_p$ are zero, and we can take r of the edges of the orthotope parallel to

r rectangular axes in the r-flat, the remaining $p - r$ being determined as above.

22. Content of the Projection of a p-Dimensional Region. The content of an orthotope with edges $a_1, \ldots, a_p$ in a p-flat α is $V = a_1 a_2 \ldots a_p$. When its projection on π is also an orthotope and α meets π in no more than one point, the content of the projection is

$$V' = V \cos \theta_1 \cos \theta_2 \ldots \cos \theta_p ;$$

and if α cuts π in an r-flat

$$V' = V \cos \theta_1 \cos \theta_2 \ldots \cos \theta_{p-r},$$

where $\theta_1, \theta_2, \ldots$ are the angles between α and π.

As any p-dimensional region can be divided into small orthotopes whose total content is in the limit equal to that of the given region, these relations hold also generally.

REFERENCE

JORDAN, C. Essai sur la géométrie à *n* dimensions. Paris, Bull. Soc. Math., **3** (1875), 103-174; **4** (1877), 92. (Abstract in C.-R. Acad. Sc., **75**, 1614-1617.)

CHAPTER V

ANALYTICAL GEOMETRY: PROJECTIVE

1. The axioms of plane projective geometry are verified when a point is represented by a set of three numbers (x, y, z), real or complex but not all zero, taken in a given order, with the understanding that if k is any factor, not zero, the set of numbers (kx, ky, kz) represents always the same point; and a line is the assemblage of points represented by numbers (x, y, z) which satisfy a fixed homogeneous equation of the first degree $lx + my + nz = 0$.

In projective geometry of n dimensions a point x is represented by the ratios of $n + 1$ numbers or co-ordinates $x_0, x_1, \ldots, x_n$, and has thus n degrees of freedom.

2. Parametric Equations. A line is a series of points, having one degree of freedom and therefore depending on a single parameter, which is completely determined by two points. We may therefore define a line as the assemblage of points represented by the equations

$$\rho x_r = x'_r + \lambda x''_r \quad (r = 0, 1, \ldots, n) \qquad . \quad (2\cdot1)$$

where x'_r and x''_r are fixed, ρ is a factor of proportionality, and λ is a variable parameter, or by

$$\rho x_r = tx'_r + ux''_r \quad (r = 0, 1, \ldots, n) \qquad . \quad (2\cdot2)$$

where the ratio t/u is the variable parameter. x' and x'' are two fixed points on the line, and correspond respectively to the values $u = 0$ and $t = 0$ of the *homogeneous parameters*.

Similarly

$$\rho x_r = tx'_r + ux''_r + vx'''_r \quad (r = 0, 1, \ldots, n) \quad . \quad (2\cdot3)$$

represents a plane through the three points x', x'', x''', provided they are not collinear. All points on the line through these points lie in the plane.

And generally

$$\rho x_r = t_1 x_r^{(1)} + \ldots + t_p x_r^{(p)} \quad (r = 0, 1, \ldots, n) \qquad (2\cdot4)$$

represents a $(p - 1)$-flat through the points $x^{(1)}, \ldots, x^{(p)}$.

3. Equation of a Hyperplane. When $p = n$ we can eliminate the $(n + 1)$ quantities $\rho, t_1, \ldots, t_n$ between these homogeneous equations, and we obtain a single equation

$$\begin{vmatrix} x_0 & x_1 & \ldots & x_n \\ x_0^{(1)} & x_1^{(1)} & \ldots & x_n^{(1)} \\ \cdot & \cdot & \cdot & \cdot \\ x_0^{(n)} & x_1^{(n)} & \ldots & x_n^{(n)} \end{vmatrix} = 0 \qquad (3\cdot1)$$

which is homogeneous and of the first degree in x_r. An $(n - 1)$-flat or hyperplane is thus represented also by a single equation of the form

$$\sum_{r=0}^{n} l_r x_r = 0 \qquad (3\cdot2)$$

4. p distinct points in general determine a $(p - 1)$-flat. The equation $(3\cdot1)$ is the condition that the $n + 1$ points $x, x^{(1)}, \ldots, x^{(n)}$ should be in the same $(n - 1)$-flat. Similarly the condition that the $p + 1$ points $x, x^{(1)}, \ldots, x^{(p)}$ should lie in the same $(p - 1)$-flat is found by eliminating the $(p + 1)$ quantities $\rho, t_1, \ldots, t_p$ between the $n + 1$ equations $(2\cdot4)$. This is represented by

$$\begin{Vmatrix} x_0 & x_1 & \ldots & x_n \\ x_0^{(1)} & x_1^{(1)} & \ldots & x_n^{(1)} \\ \cdot & \cdot & \cdot & \cdot \\ x_0^{(p)} & x_1^{(p)} & \ldots & x_n^{(p)} \end{Vmatrix} = 0 \qquad (4\cdot1)$$

which implies the vanishing of every determinant obtained by striking out any $n - p$ columns of this array or matrix. As a point in S_n has n degrees of freedom, and in S_{p-1} has $p - 1$ degrees of freedom, a point constrained to lie in the S_{p-1} determined by p given points has $n - p + 1$ degrees of constraint. The equation $(4\cdot1)$ which involves the vanishing of ${}_{n+1}C_{p+1}$ determinants, is equivalent to only $n - p + 1$ conditions.

5. Matrix Notation. As we shall frequently have to use this notation it is necessary to state some definitions and theorems regarding matrices.

A matrix is a rectangular arrangement of mn elements a_{mn} in m rows and n columns, which may be denoted by $[a_{mn}]$. If we choose any r rows and the same number of columns, an r-rowed determinant is formed which is called a determinant of the matrix. If every $(r + 1)$-rowed determinant of a matrix vanishes, but there is at least one r-rowed determinant which does not vanish, the matrix is said to be of *rank r.*

6. The vanishing of all the r-rowed determinants of the matrix $[a_{mn}]$ will be denoted by the equation

$$\| a_{mn} \|_r = 0,$$

or, more accurately, by the statement that the matrix is of rank $r - 1$.

The number of r-rowed determinants in this matrix is ${}_mC_r \cdot {}_nC_r$, but they are not all independent, and if a certain number of them, suitably chosen, are equated to zero, the rest will vanish identically. For example take the matrix

$$\begin{bmatrix} a_1 & b_1 & c_1 \\ a_2 & b_2 & c_2 \end{bmatrix}.$$

If $a_1b_2 - a_2b_1 = 0$ and $a_1c_2 - a_2c_1 = 0$, then, provided a_1 and a_2 do not both vanish, we have

$$\frac{b_1}{b_2} = \frac{a_1}{a_2} = \frac{c_1}{c_2}$$

and therefore $b_1c_2 - b_2c_1 = 0$.

THEOREM. *The number of independent conditions involved in the equation*

$$\| a_{mn} \|_r = 0$$

is in general $(m - r + 1)(n - r + 1)$.

Consider first the array

$$\| a_{rn} \|_r,$$

and let all the determinants $(n - r + 1$ in number) vanish which have the same $r - 1$ columns, only differing, say, in the last column; expanding each in terms of the co-factors

$A_{\mu\nu} = A_{\mu\nu'} = A_{\mu}$, say, of the elements of the last column, we have

$$\sum_{\mu=1}^{r} a_{\mu\nu}A_{\mu} = 0 \ (\nu = r, r+1, \ldots, n).$$

We assume that the A's are not all zero. Now take any other determinant of the array

$$\begin{vmatrix} a_{1\nu_1} & a_{1\nu_2} & \ldots & a_{1\nu_r} \\ \cdot & \cdot & & \cdot \\ a_{r\nu_1} & a_{r\nu_2} & \ldots & a_{r\nu_r} \end{vmatrix}.$$

Multiply the rows respectively by $A_1, A_2, \ldots, A_r$, and add to the first row. In the pth column we have then as element of the first row

$$\sum_{\mu=1}^{r} a_{\mu\nu_p} A_{\mu}.$$

If $\nu_p = 1, 2, \ldots$ or $r - 1$ this is a determinant which has two columns the same and therefore vanishes identically, and if $\nu_p = r, r+1, \ldots$ or n, it is one of the determinants which already vanish. Hence all the elements of the first row vanish, and the whole determinant vanishes identically.

Taking now the array $\| a_{mn} \|_r$, if we equate to zero all the r-rowed determinants which differ only in the elements of the last column or the last row (the number of these being $(m - r + 1)(n - r + 1)$), the previous proof shows that all the determinants vanish which have their first $r - 1$ rows chosen from the elements of the first $r - 1$ rows of the array, and then applying the same result to columns instead of rows we deduce the vanishing of all the other determinants of the array as well.

7. Linearly Independent Points. If p points do not all lie in the same $(p - 2)$-flat it follows that no set of r of them lie in the same $(r - 2)$-flat, for $r = 2, 3, \ldots, p$. The p points are then said to be *linearly independent.* The condition for this can be expressed by saying that the matrix

$$\begin{bmatrix} x_0^{(1)} & \ldots & x_n^{(1)} \\ \cdot & \cdot & \cdot \\ x_0^{(p)} & \ldots & x_n^{(p)} \end{bmatrix}$$

is of rank p. If the matrix is of rank $p - 1$, the p points all lie in the same $(p - 1)$-flat, but not in a $(p - 2)$-flat. Similarly it can be shown that the necessary and sufficient condition that the p points should all lie in the same q-flat, but not in the same $(q - 1)$-flat is that the matrix is of rank q.

8. p hyperplanes $(p \gtreqless n)$ in general have in common an $(n - p)$-flat, and $n + 1$ hyperplanes have in general no point in common. If no set of r of the p hyperplanes have in common an $(n - r + 1)$-flat $(r = 3, 4, \ldots, p)$ they are said to be *linearly independent*. If their equations are $\Sigma\xi_r^{(s)}x_r = 0$ $(s = 1, \ldots, p)$ the condition for this is that the matrix

$$\begin{bmatrix} \xi_0^{(1)} & \cdots & \xi_n^{(1)} \\ \cdot & \cdot & \cdot \\ \xi_0^{(p)} & \cdots & \xi_n^{(p)} \end{bmatrix}$$

is of rank p. If the matrix is of rank q $(< p)$, the p hyperplanes all contain the same $(n - q)$-flat. This holds also if $p > n$.

Clearly not more than $n + 1$ points or hyperplanes can be independent since the matrix has only $n + 1$ columns and cannot be of higher rank than $n + 1$.

9. Simplex of Reference. The $n + 1$ points $A_0 \equiv (1, 0, \ldots, 0)$, $A_1 \equiv (0, 1, 0, \ldots, 0)$, $\ldots$, $A_n \equiv (0, 0, \ldots, 0, 1)$ are independent and are called the points of reference or fundamental points of the co-ordinate-system. They form a simplex called the simplex of reference or fundamental simplex. The equations of its $(n - 1)$-dimensional boundaries or fundamental hyperplanes are $x_0 = 0, x_1 = 0, \ldots, x_n = 0$.

10. Unit-point; Cross-ratios. To fix the co-ordinate-system it is not sufficient to fix the points of reference. We are still free to give any values to the co-ordinates of a particular point; and we shall choose a certain point, not on any of the fundamental hyperplanes, and give its co-ordinates all the value unity. We call this the *unit-point*, $U \equiv (1, 1, \ldots, 1)$.

It can now be shown that the co-ordinates of any other point are determined, in terms of certain cross-ratios. Let $P \equiv (x_0', \ldots, x_n')$, join UP and let it cut the hyperplane opposite A_r in L_r. Let the cross-ratio $(L_0L_r, UP) = X_r$. The co-ordinates of any point on UP are given by

$$\rho x_\nu = tx_\nu' + u \quad (\nu = 0, \ldots, n)$$

where t, u are homogeneous parameters. By a fundamental theorem (which we assume) since there is a (1, 1) correspondence between the points of the line and the ratios of the homogeneous parameters t/u ($= \lambda$), the cross-ratio of four points on the line is equal to the cross-ratio of the parameters, this cross-ratio being defined to be

$$(\lambda_1\lambda_2, \lambda_3\lambda_4) \equiv \frac{\lambda_1 - \lambda_3}{\lambda_2 - \lambda_3} \Big/ \frac{\lambda_1 - \lambda_4}{\lambda_2 - \lambda_4}.$$

The parameters of the four points L_0, L_r, U, P are given by $0 = x'_0 t_1 + u_1$, $0 = x'_r t_2 + u_2$, $t_3 = 0$, $u_4 = 0$, and the cross-ratio

$$(\lambda_1\lambda_2, \lambda_3\lambda_4) = \frac{-x'_r}{-x'_0}.$$

Hence $$\frac{x'_r}{x'_0} = X_r.$$

11. Duality. A point and an $(n - 1)$-flat in S_n have the same number of degrees of freedom and can be put into (1, 1) correspondence. The assemblage of all $(n - 1)$-flats in S_n is an n-dimensional manifold or space of n dimensions which we may denote by Σ_n. The $(n - 1)$-flat is here regarded as the element and can be fixed by co-ordinates like a point. The simplest correspondence between S_n and Σ_n is established by defining the homogeneous co-ordinates of the $(n - 1)$-flat whose point-equation is $\Sigma\xi_r x_r = 0$ to be $(\xi_0, \xi_1, \ldots, \xi_n)$.

There is an exact duality between points and $(n - 1)$-flats, lines and $(n - 2)$-flats, and so on. To an $(n - 1)$-flat considered as an assemblage of points corresponds an assemblage of $(n - 1)$-flats through a point. The equations

$$\rho\xi_r = \xi'_r + \lambda\xi''_r \quad (r = 0, 1, \ldots, n)$$

represent an assemblage of $(n - 1)$-flats all passing through the $(n - 2)$-flat in which ξ' and ξ'' intersect.

12. Order of a Variety. A homogeneous equation of degree r in the point-co-ordinates x_ν represents an $(n - 1)$-dimensional assemblage of points or a *variety of order* r, characterised by the property that it is cut by an arbitrary line in r points, real, coincident, or imaginary. It is denoted by V^r_{n-1}. A system of p homogeneous equations in x_ν represents a variety of $n - p$ dimensions V^r_{n-p}, the intersection of p varieties of $n - 1$ dimensions. Its order r is equal to the number of points in which it is cut by an arbitrary p-flat.

13. A system of equations of the form

$$\rho x_\nu = \phi_\nu(\lambda) \quad (\nu = 0, 1, \ldots, n),$$

where ϕ_ν denote functions of the single parameter λ, represents a curve. If the ϕ's are rational, integral, algebraic functions, the curve is said to be *rational.* If the functions are each of degree r the order of the curve is r, for it is cut by an arbitrary $(n - 1)$-flat $\Sigma l_\nu x_\nu = 0$ in r points corresponding to the r roots of the equation $\Sigma l_\nu \phi_\nu = 0$.

A system of equations of the form

$$\rho x_\nu = \phi_\nu(t_0, t_1, \ldots, t_p) \quad (\nu = 0, 1, \ldots, n),$$

where ϕ_ν denote rational, integral, algebraic functions, homogeneous in the $p + 1$ parameters t, each of degree r, represents a *rational* V_p.

Generally, a V^r_p is a variety of p dimensions of order r, i.e. such that an arbitrary $(n - p)$-flat in the S_n which contains it cuts it in r points.

14. **Rational Normal Curves.** *A curve of order r lies entirely in a flat space of dimensions r or less*, for if $r + 1$ arbitrary points are taken on the curve they determine an S_r which would cut the curve in $r + 1$ points, and must therefore contain the whole curve.*

A curve of order n in S_n is always rational, for if we take $n - 1$ fixed points on the curve there is a one-dimensional pencil of $(n - 1)$-flats through these fixed points, and each cuts the curve in one other point. There is thus a $(1, 1)$ correspondence between the points of the curve and the values of the parameter which determines hyperplanes of the pencil. A curve of order

* Or a portion of the curve. In the latter case the curve is not a proper curve of order r, but decomposes. Thus a proper cubic curve must be contained in a space of three dimensions; but a conic together with a line which does not meet the plane of the conic form a composite cubic which is not contained in a space of fewer than four dimensions. Three mutually skew lines form a composite cubic which may be contained in a space of not fewer than five dimensions.

r which is not contained in any flat space of fewer than r dimensions is called a rational *normal* curve; e.g. a conic, a twisted cubic, a curve of the fourth order in S_4, and so on.

15. *A normal curve of order r cannot be cut by any line in more than two points*, for suppose a certain line cuts the curve in three points, then the line together with $r - 2$ other points on the curve determines an $(r - 1)$-flat which cuts the curve in $r + 1$ points; and, generally, *a curve of order r which is contained in a flat space of n dimensions* ($n \lesseqgtr r$) *cannot be cut by a p-flat* ($p < n$) *in more than* $p + r - n + 1$ *points*, for suppose it to be cut in $p + r - n + 2$ points; then through the p-flat, which is determined by $p + 1$ points, and $n - p - 1$ other points on the curve there passes an $(n - 1)$-flat which cuts the curve in $(p + r - n + 2) + (n - p - 1) = r + 1$ points.

A surface of order r, i.e. a V_2^r, in S_n, but not lying in an S_{n-1} is cut by an arbitrary S_{n-2} in r points. The S_{n-2} is determined by $n - 1$ points. Hence $n - 1 \lesseqgtr r$, i.e. *a* V_2^r *is always contained in a flat space of* $r + 1$ *dimensions or less.*

A V_p^r in S_n, but not lying in an S_{n-1}, is cut by an arbitrary S_{n-p} in r points. The S_{n-p} is determined by $n - p + 1$ points. Hence $n - p + 1 \lesseqgtr r$, i.e. *a* V_p^r *is always contained in a flat space of* $r + p - 1$ *dimensions or less.* In particular a V_p^2, or quadric variety, is always contained in a flat space of $p + 1$ dimensions.

A V_q^r *which is contained in a flat space of n dimensions* ($n \lesseqgtr r + q - 1$) *cannot be cut by an* S_p ($p \lesseqgtr n - q$) *in more than* $p + r - n + q$ *points.*

16. Rational Normal Varieties. A V_{n-r+1}^r which is contained in a space of dimensions n but not fewer is said to be *normal*, and is rational. For if we take $r - 1$ fixed points on the variety there is an $(n - r + 1)$-dimensional linear system of $(r - 1)$-flats through these $r - 1$ points, and each cuts the variety in one other point. Thus the co-ordinates of this variable point can be expressed rationally in terms of $n - r + 1$ variable parameters; or the points of the variety can be put into (1, 1) correspondence with the points of a flat space of $n - r + 1$ dimensions. A quadric variety of any dimensions is rational. A quadric in space of three dimensions can be represented rationally on a plane; the process, already indicated in the general proof, is by projecting the surface on to the plane from a centre of projection lying on the surface, i.e. stereographic projection.

In a plane the only rational normal variety is the conic. In three dimensions the only rational normal varieties are the cubic curve, and the quadric surface. In four dimensions we have the quartic curve, the cubic surface, and the 3-dimensional quadric. In S_n the only rational normal varieties are the V_1^n, $V_2^{n-1}, \ldots, V_{n-1}^2$. There are of course other *rational* varieties, e.g. rational plane cubic curves, rational cubic surfaces in S_3, etc., but for such varieties special conditions have to be satisfied.

Quadric Varieties

17. We shall consider now in more detail the quadric variety, represented by a homogeneous equation in x_r of the second degree

$$\sum_{r=0}^{n}\sum_{s=0}^{n} a_{rs}x_r x_s = 0 \quad . \quad . \quad . \quad (17\cdot 1)$$

When $s = r$ we have a "square" term $a_{rr}x_r^2$; when $s \neq r$ we have "product" terms, and the coefficient of $x_r x_s$ is $a_{rs} + a_{sr}$; this can be written as a single coefficient, say $2a'_{rs}$, which is equivalent to assuming $a_{rs} = a_{sr}$. This assumption therefore does not lead to any loss of generality.

18. Conjugate Points, Polar, Tangent. Let α and β be any two points, then any point on their join is represented by

$$\rho x_\nu = \lambda\alpha_\nu + \mu\beta_\nu \; (\nu = 0, 1, \ldots, n) \quad . \quad (18\cdot 1)$$

If x lies on the quadric we have

$$\Sigma\Sigma a_{rs}(\lambda\alpha_r + \mu\beta_r)(\lambda\alpha_s + \mu\beta_s) = 0,$$

i.e. $$\lambda^2\Sigma\Sigma a_{rs}\alpha_r\alpha_s + 2\lambda\mu\Sigma\Sigma a_{rs}\alpha_r\beta_s + \mu^2\Sigma\Sigma a_{rs}\beta_r\beta_s = 0 \quad (18\cdot 2)$$

This quadratic equation in $\lambda : \mu$ determines the two points in which the line joining α and β cuts the quadric. If these points are harmonic conjugates with regard to α and β the two values of $\lambda : \mu$ must be equal but opposite in sign, and therefore

$$\Sigma\Sigma a_{rs}\alpha_r\beta_s = 0 \quad . \quad . \quad . \quad (18\cdot 3)$$

The two points α, β are then said to be *conjugate* with regard to the quadric. If α is fixed, the locus of harmonic conjugates, or the *polar* of α is

$$\Sigma\Sigma a_{rs}\alpha_r x_s = 0 \quad . \quad . \quad . \quad (18\cdot 4)$$

If α and β are conjugate points, and α lies on the quadric, equation (18·2) becomes

$$\mu^2 \Sigma\Sigma a_{rs}\beta_r\beta_s = 0 \quad . \quad . \quad . \quad (18\cdot 5)$$

If $\mu \neq 0$ then $\Sigma\Sigma a_{rs}\beta_r\beta_s = 0$; in this case β lies on the quadric, the value of $\lambda : \mu$ is indeterminate, and all points of the line $\alpha\beta$ lie on the quadric; we shall return to this case later. If β does not lie on the quadric, $\mu = 0$, and the point of intersection x coincides with α. In this case the line $\alpha\beta$ cuts the quadric in two coincident points at α and is said to be a *tangent* to the quadric; if α is fixed and β variable the locus of β consists of all points on all the tangents through α, this is the *tangent hyperplane* at α; if β is fixed and α variable the locus of α consists of all the points of contact of tangents through β. In the former case, when α and β are conjugate and both lie on the quadric, every point on the line $\alpha\beta$ also lies on the quadric.

19. Correlations. The correspondence between a point and its polar hyperplane with regard to a quadric variety is a special case of the general (1, 1) correspondence or correlation between points and hyperplanes; such a correlation is possible since the point and the hyperplane have the same number of degrees of freedom. The assemblage of all points, and the assemblage of all hyperplanes, are manifolds of the same dimensions. If the (1, 1) correspondence is determined by the equations

$$\xi'_r = \sum_{s=0}^{n} a_{rs}x_s \quad (r = 0, 1, \ldots, n) \quad . \quad (19\cdot 1)$$

where x_s are the homogeneous co-ordinates of the point, and ξ'_r the coefficients in the equation of the hyperplane, or the homogeneous co-ordinates of the hyperplane (we do not assume here that $a_{rs} = a_{sr}$), the assemblage of points which lie on the corresponding hyperplane is represented by the equation $\Sigma x_r\xi'_r = 0$, i.e.

$$\Sigma\Sigma a_{rs}x_r x_s = 0, \quad . \quad . \quad . \quad (19\cdot 2)$$

a quadric-locus F.

If the equations (1) are solved for x we get

$$x_r = \Sigma A_{sr}\xi'_s \quad (r = 0, 1, \ldots, n) \qquad . \quad (19{\cdot}3)$$

where A_{rs} is the cofactor of a_{rs} in the determinant $| a_{nn} | \equiv D$. It is assumed that this determinant does not vanish, so that

$$\sum_{s=0}^{n} a_{rs}A_{rs} = D, \text{ and } \sum_{s=0}^{n} a_{rs}A_{ts} = 0 \text{ for } r \neq t.$$

If $D = 0$ the hyperplanes $\Sigma a_{rs}x_s = 0$ have at least one point common to all, and every hyperplane in the S-space which corresponds to a point in the S′-space passes through this fixed point. The correlation is then said to be *singular*. We exclude this case for the present.

From equation (19·3) we obtain similarly the assemblage of hyperplanes which pass through the corresponding point, viz.

$$\Sigma\Sigma A_{rs}\xi_r\xi_s = 0, \quad . \quad . \quad . \quad (19{\cdot}4)$$

a quadric-envelope Φ. F and Φ are in general different figures. If x is a point of F the corresponding hyperplane is ξ', which is a tangent hyperplane to Φ, but not in general to F. To the point x' in the space S′ corresponds a hyperplane. The point x' is represented in the space S′ by the equation

$$\Sigma x'_r\xi'_r = 0,$$

i.e.

$$\Sigma\Sigma a_{rs}x'_r x_s = 0. \quad . \quad . \quad . \quad (19{\cdot}5)$$

But this is the hyperplane whose co-ordinates are

$$\xi_s = \Sigma a_{rs}x'_r \quad \text{or} \quad \xi_r = \Sigma a_{sr}x'_s \quad . \quad . \quad (19{\cdot}6)$$

20. Polar System, Null System. If the points x and x' which correspond to the same hyperplane (in S′ and S respectively) are the same, we see from (19·1) and (19·6) that the equations

$$\Sigma a_{rs}x_s = 0 \quad \text{and} \quad \Sigma a_{sr}x_s = 0$$

must be identical, therefore

$$a_{rs} = \rho a_{sr}.$$

Interchanging r and s, we have

$$a_{sr} = \rho a_{rs}.$$

Hence

$$\rho^2 = 1 \text{ and } \rho = \pm 1.$$

If $\rho = +1$ we have then the conditions

$$a_{rs} = a_{sr}.$$

The correlation is symmetrical and is a *polar system* with respect to the quadric

$$\Sigma\Sigma a_{rs}x_r x_s = 0.$$

In this case the quadrics F and Φ are the same figure, as locus and envelope respectively.

If $\rho = -1$ we have

$$a_{rs} = -a_{sr}, \quad a_{rr} = 0.$$

In this case the quadrics F and Φ do not exist, since their coefficients all vanish identically. This is called a *null system.* Every point lies on its corresponding hyperplane and *vice versa.* A non-singular null system cannot, however, exist in space of even dimensions, for in that case the determinant D, being a skew symmetric determinant of odd order, vanishes identically.

21. Canonical Equation of a Quadric. In the polar system, if the simplex of reference is chosen so that each vertex is the pole of the opposite hyperplane, the equations of the polarity (19·1) become

$$\xi'_r = a_r x_r$$

and the equation of the quadric (19·2) is

$$\Sigma a_r x_r^2 = 0.$$

This is called its *canonical* equation, referred to a self-conjugate simplex.

22. The Real Lines, Planes, etc., which lie on a Quadric. If the two distinct points α and β are conjugate with regard to the quadric and if both of them lie on the quadric we have seen that every point on their join also lies on the quadric. We have then a line $\alpha\beta$ lying entirely on the quadric. If γ is a third point also lying on the quadric, conjugate to both α and β, but not lying on the line $\alpha\beta$, it follows that every point of the plane $\alpha\beta\gamma$ will lie entirely on the quadric, and so on.

Consider the quadric whose equation is

$$a_0x_0^2 + \ldots + a_{k-1}x_{k-1}^2 - a_kx_k^2 - \ldots - a_nx_n^2 = 0,$$

where the a's are all positive, i.e. there are k positive terms and $n - k + 1$ negative terms. We assume that $k \lessgtr \frac{1}{2}n$; if this is not so we may change the signs all through and then $a_k, \ldots, a_n$ become the positive coefficients. The edge A_0A_k of the fundamental simplex, which has the equations

$$x_1 = 0, \ldots, x_{k-1} = 0, x_{k+1} = 0, \ldots, x_n = 0$$

cuts the quadric in two real points P_0, Q_0 say, determined by $a_0x_0^2 - a_kx_k^2 = 0$. Similarly A_1A_{k+1}, A_2A_{k+2}, ..., $A_{k-1}A_{2k-1}$ cut the quadric in the pairs of points

$$P_1, Q_1; \; P_2, Q_2; \; \ldots; \; P_{k-1}, Q_{k-1}.$$

Now P_0 and Q_0 are conjugate to all points lying on the hyperplane $A_1A_2 \ldots A_{k-1}A_{k+1} \ldots A_n$, and therefore to all the other points P_r, Q_r, and similarly any two of the $2k$ points, with different suffixes, are conjugate. Further, any k of them whose suffixes are all different are independent, i.e. do not lie in the same $(k - 2)$-flat; the $(k - 2)$-flat determined by

$$P_0, P_1, \ldots, P_{k-2}$$

e.g. does not cut the edge $A_{k-1}A_{2k-1}$. Hence the k points P_0, P_1, ..., P_{k-1} are independent and conjugate in pairs, and hence the $(k - 1)$-flat determined by them lies entirely in the quadric. Now no real flat of higher dimensions can lie entirely in the quadric, for it would cut in at least one point every $(n - k)$-flat, e.g. the flat $A_kA_{k+1} \ldots A_n$; but the section of the quadric by this flat is

$$a_kx_k^2 + \ldots + a_nx_n^2 = 0$$

which contains no real points.

Hence *if the canonical equation of the quadric contains k terms of one sign and k' terms of opposite sign ($k + k'$ being $=$ $n + 1$), and if $k \lessgtr k'$, the quadric contains real flats of dimension $k - 1$ and no higher dimension.*

23. If the equation $\Sigma\Sigma a_{rs}x_rx_s = 0$ is transformed by any linear equations with real coefficients into $\Sigma\Sigma a'_{rs}x'_rx'_s = 0$, and if the canonical forms of the equations are

$$\Sigma a_ry_r^2 = 0, \qquad \Sigma a'_ry_r'^2 = 0$$

then the two expressions on the left-hand sides have the same number of positive coefficients. This is known as Sylvester's *Law of Inertia.*

24. Specialised Quadrics. When the equation of the quadric is reduced to its canonical form

$$\Sigma a_r x_r^2 = 0$$

the nature of the quadric projectively, i.e. apart from metrical properties, depends only on the signs of the coefficients a_r. If the coefficients are all of the same sign the quadric contains no real points and is said to be *virtual.* We have so far disregarded the cases in which one or more of the coefficients are zero, and excluded the case in which the determinant D vanishes. But when the equation is in the canonical form the determinant D reduces to the product $a_0 a_1 a_2 \ldots a_n$, so that when $D = 0$, one or more of the coefficients vanish. When r of the coefficients vanish the quadric is said to be specialised r times. The condition for this is that the matrix or determinant D is of rank $n - r + 1$.

If one of the coefficients, a_0, vanishes, the polar hyperplane of the point y is

$$a_1 y_1 x_1 + a_2 y_2 x_2 + \ldots + a_n y_n x_n = 0$$

and this always passes through the point $A_0 \equiv (1, 0, \ldots, 0)$; and the polar hyperplane of A_0 is quite indeterminate. Any line through A_0,

$$\rho x_0 = t x_0' + 1,$$
$$\rho x_r = t x_r', \qquad (r = 1, \ldots, n)$$

cuts the quadric where

$$\sum_{r=1}^{n} t^2 a_r x_r' = 0.$$

This equation is satisfied for all values of t if x' lies on the quadric, but gives equal roots $t^2 = 0$ for other lines. Hence A_0 is a *double-point* on the quadric. The quadric is a *hypercone* with vertex A_0 and can be generated by a line which passes always through A_0. Similarly if two coefficients vanish, a_0 and a_1, the polar hyperplanes of all points contain the line A_0A_1

while the polars of points on this line are indeterminate. A_0A_1 is a *double-line*, and the quadric can be generated by planes which pass through A_0A_1. In this case the quadric is said to be specialised twice and is called a hypercone of the second species; A_0A_1 is called the vertex-edge. And in general when r coefficients vanish the quadric is specialised r times and is a hypercone of species r with an $(r - 1)$-flat as vertex-edge.

25. Hypercones. A hypercone can be defined more generally as follows. If V is a variety of $n - 2$ dimensions, and O a point not lying in its hyperplane, the ruled hypersurface generated by straight lines joining O to the points of V is a hypercone of $n - 1$ dimensions. V is the base and O the vertex. Any section of the hypercone by a hyperplane not passing through O is a variety of the same order as V and may equally serve as base. We may call it more properly *director*. A section by a hyperplane passing through O is a hypercone of $n - 2$ dimensions.

To find the equation of a hypercone with vertex at $A_0 \equiv (1, 0, \ldots, 0)$ and base a variety

$$u_0x_0^r + u_1x_0^{r-1} + u_2x_0^{r-2} + \ldots = 0, \quad \alpha_0x_0 + \alpha_1 = 0,$$

where u_r and α_r denote homogeneous polynomials of degree r in $x_1, \ldots, x_n$.

Let y be any point on the base. Then the freedom-equations of a line joining A_0 to y are

$$\rho x_\nu = y_\nu \quad (\nu = 1, \ldots, n)$$
$$\rho x_0 = t.$$

Then
$$t\beta_0 + \beta_1 = 0,$$
$$t^r v_0 + t^{r-1}v_1 + t^{r-2}v_2 + \ldots = 0,$$

where $\beta_0, \beta_1, v_0, v_1, \ldots$ are the same functions of y as $\alpha_0, \alpha_1, u_0, u_1, \ldots$ are of x. Eliminating t we get

$$\beta_1^r v_0 - \beta_1^{r-1}\alpha_0 v_1 + \beta_1^{r-2}\alpha_0^2 v_2 - \ldots = 0.$$

This equation is homogeneous in $(x_1, \ldots, x_n)$.

If the base V is itself a hypercone of $n - 2$ dimensions with vertex A_0 the hypercone generated by taking a vertex A_1 is called a *hypercone of the second species*. Every section of this cone by a hyperplane not passing through both A_0 and A_1 is a hypercone and could equally well be taken as base, and then any point on the line A_0A_1 as vertex. The vertices A_0 and A_1 are therefore not specialised in any way, but are any two points on a certain fixed line, the *vertex-edge*. If V′ is an $(n - 3)$-dimensional section of the base V, not passing through the vertex, the hypercone with edge A_0A_1 can be generated by the planes which pass through the line A_0A_1 and points of V′. A hypercone of the second species is thus determined by a straight line as vertex-edge and a variety of $n - 3$ dimensions as director. Any section not containing the vertex-edge cuts the hypercone of the second species in a hypercone of the first species, and any section containing the vertex-edge cuts it in a hypercone of the second species.

Generally, a hypercone of $n - 1$ dimensions and of species k is generated by the k-flats which pass through a given $(k - 1)$-flat, the vertex-edge, and points of a given variety V_{n-k-1}. Any section by a hyperplane which cuts the vertex-edge in a p-flat is a hypercone of species $p + 1$.

For a hypercone of species k in S_n, with vertex-edge a $(k - 1)$-flat whose equations are $x_0 = 0, x_1 = 0, \ldots, x_{n-k} = 0$, and director a variety V_{n-k-1} whose equations are $u_1 = 0$, $u_2 = 0, \ldots, u_{k+1} = 0$, eliminating the k co-ordinates x_{n-k+1}, $\ldots, x_n$ between the $k + 1$ equations $u_\nu = 0$ we get the equation of the hypercone, which is homogeneous in $x_0, x_1, \ldots$, x_{n-k}.

26. Polar Spaces. Since the relation between conjugate points is symmetrical it follows that if the polar hyperplane of a point P passes through Q, then the polar hyperplane of Q will pass through P. The polars of P and Q intersect in an S_{n-2}, and the polar of any point on S_{n-2} passes through both P and Q and therefore contains the line $PQ \equiv S_1$, and the polars of all points on S_1 contain S_{n-2}; S_1 and S_{n-2} are called polars with regard to the quadric. More generally, the polars of all points on an S_p pass through an S_{n-p-1} and vice versa; S_p and S_{n-p-1} are called polars with regard to the quadric. Every point of S_p is conjugate to every point of its polar S_{n-p-1}. S_p cuts the quadric in a V^2_{p-1}, and the tangent hyperplanes to V^2_{p-1} at the points of section all pass through the polar S_{n-p-1}.

27. Tangent Spaces. An S_p and its polar S_{n-p-1} in general have no point in common. If they intersect and have in common an S_r, the points of S_r are all self-conjugate and therefore S_r lies entirely in the quadric; these points are also conjugate to all points of S_p and S_{n-p-1} and therefore the section V^2_{p-1} of the quadric by S_p (or V^2_{n-p-2} by S_{n-p-1}) is a hypercone having S_r as vertex-edge. In particular the tangent hyperplane at any point P, which is the polar of the point P, cuts the quadric in a hypercone of the first species with P as vertex (it may of course be virtual with P as its only real point). When S_p and its polar S_{n-p-1} have an S_r in common, S_p is said to be a tangent to the quadric in S_r or in any linear space which is contained in S_r. This includes the case of a point being tangent to a quadric when it lies on the quadric. An S_{n-1} can only touch in a point (unless the quadric is specialised as a hypercone), an S_{n-2} can touch in a point or a line, and so on.

If S_p lies entirely in the quadric, each of its points is conjugate to all the points of any linear space which touches the quadric in S_p or in a linear space containing S_p. Hence the polar S_{n-p-1} of S_p contains not only S_p but also every linear space containing S_p and lying in the quadric.

28. Conditions for a Hypercone. If the quadric

$$\Sigma\Sigma a_{rs}x_r x_s = 0 \quad . \quad . \quad . \quad (29{\cdot}1)$$

is specialised k times, so that it is a hypercone of species k, with a double $(k-1)$-flat S_{k-1} as vertex, the polar hyperplane of every point contains S_{k-1}. Taking the polars of $A_0, A_1, \ldots, A_n$, it follows that the $n+1$ hyperplanes

$$\Sigma a_{rs}x_r = 0 \quad (s = 0, 1, \ldots, n) \quad . \quad (29{\cdot}2)$$

all pass through the same S_{k-1}. The condition for this is that the matrix $\| a_{nn} \|$ should be of rank $n-k+1$. Then any $n-k+1$ of the equations (29·2) determine the vertex-edge.

29. Degenerate Quadrics. In a plane S_2 there are no proper cones. Considering cones of the second order only, a cone of the first species in S_2 degenerates to two lines through a point. In S_3 a cone of the second species degenerates to two planes through a line. In S_n a hypercone of species $n-1$

degenerates to two $(n-1)$-flats through an $(n-2)$-flat; a hypercone of species n would have an $(n-1)$-flat as vertex or double-space, and must therefore be considered as degenerating to this $(n-1)$-flat counted twice.

In S_{2n} if an n-flat S_n cuts its polar S_{n-1} with regard to a given V^2_{2n-1} in an S_{n-2}, S_n touches V^2_{2n-1} at the S_{n-2} and meets the quadric in a hypercone V^2_{n-1} of species $n-1$ which therefore consists of two $(n-1)$-flats; if S_n contains its polar S_{n-1}, it touches the quadric in this S_{n-1} and meets the quadric in this S_{n-1} twice.

In S_{2n+1} if an S_{n+1} contains its polar S_{n-1} with regard to a given V^2_{2n} it touches the quadric in this S_{n-1}, and the section consists of two n-flats; if an S_n cuts its polar S_n in an S_{n-1} it touches the quadric in this S_{n-1} and the section is this S_{n-1} counted twice; if an S_n coincides with its polar S_n it lies entirely in the quadric.

30. Linear Spaces on a Quadric. We shall consider now in more detail, and apart from the question of reality, the lines, planes, etc., lying on a quadric. Take any point C_0 on the quadric, and let $S^{(0)}_{n-1}$ be its polar; this therefore cuts the quadric in a hypercone V^2_{n-2} having C_0 as a double-point. On this hypercone take a point C_1 distinct from C_0, then the line C_0C_1 lies on the quadric; and the polar $(n-2)$-flat of C_0C_1 is a hypercone V^2_{n-3} of the second species having C_0C_1 as vertex-edge. On this hypercone take a third point C_2, not on C_0C_1, then the plane $C_0C_1C_2$ lies on the quadric, and is the vertex-edge for the hypercone V^2_{n-4} of the third species in which the polar $(n-3)$-flat of $C_0C_1C_2$ cuts the quadric, and so on.

We have now to distinguish according as n is odd or even.

If n is even, $=2p$, when we have obtained in this way p points $C_0, C_1, \ldots, C_{p-1}$, the $(p-1)$-flat determined by them lies in the quadric, and its polar p-flat cuts the quadric in a hypercone having this $(p-1)$-flat as vertex-edge; this is a degenerate hypercone consisting of the $(p-1)$-flat counted twice. This terminates the process. At any stage of course the hypercone may have no real elements except its vertex edge, and as regards real elements the process may thus terminate earlier.

If n is odd, $= 2p + 1$, when we have obtained p points C_0, $C_1, \ldots, C_{p-1}$, we again have a $(p - 1)$-flat on the quadric, and its polar is a $(p + 1)$-flat which cuts the quadric in a hypercone having the $(p - 1)$-flat as vertex-edge; this hypercone therefore degenerates to two p-flats through the vertex-edge, and the process then terminates.

Hence *a* V^2_{n-1} *in* S_n *contains linear spaces of one, two, . . . up to* $\frac{1}{2}n - 1$ *or* $\frac{1}{2}(n - 1)$ *dimensions, according as n is even or odd;* or we may say, since the quadric is of $n - 1$ dimensions, *the dimensions of a linear space contained in a quadric are equal to or less than half the dimensions of the quadric.*

31. To find the number of degrees of freedom of a p-flat in a V^2_{n-1}. The first point C_0 has $n - 1$ degrees of freedom; the second C_1 has $n - 2$; the third $n - 3$, and so on. Hence the $p + 1$ points $C_0, C_1, \ldots, C_p$ have

$$\sum_{r=1}^{p+1} (n - r) = n(p + 1) - \tfrac{1}{2}(p + 1)(p + 2).$$

But each point has p degrees of freedom in the p-flat, hence the p-flat has

$$n(p+1) - \tfrac{1}{2}(p+1)(p+2) - p(p+1) = \tfrac{1}{2}(p+1)(2n - 3p - 2)$$

degrees of freedom in the V^2_{n-1}.

To find the number of degrees of freedom of a p-flat passing through a q-flat in the V^2_{n-1}. The first $q+1$ points $C_0, C_1, \ldots, C_q$ are fixed, and the remaining points have

$$\sum_{r=q+2}^{p+1} (n - r) = n(p - q) - \tfrac{1}{2}(p - q)(p + q + 3)$$

degrees of freedom, but each of the last $p - q$ points has p degrees of freedom in the p-flat, hence the p-flat has

$$n(p-q) - \tfrac{1}{2}(p-q)(p+q+3) - p(p-q) = \tfrac{1}{2}(p-q)(2n - 3p - q - 3)$$

degrees of freedom.

If $n = 2p$, the number of degrees of freedom of a $(p - 1)$-flat, the space of highest dimensions, is $\frac{1}{2}p(p + 1) = \frac{1}{8}n(n+2)$, and there is a single infinity of $(p - 1)$-flats through every $(p - 2)$-flat.

If $n = 2p + 1$, the number of degrees of freedom of a p-flat, the space of highest dimensions, is again $\frac{1}{2}p(p + 1) = \frac{1}{8}(n^2 - 1)$, and there are two p-flats through every $(p - 1)$-flat.

32. In S_3 there are two singly infinite systems of lines on a quadric, two through every point, and every line of the one system cuts every line of the other system, but no two lines of the same system intersect.

In S_4 a quadric contains a triple infinity of lines, and through every point P a single infinity of lines forming a cone with vertex P in the tangent hyperplane at P. The lines through any point therefore form one continuous series and are not separated into two groups. Any other line of the quadric cuts the tangent hyperplane at P in a point Q which also lies on the quadric and therefore cuts one of the lines which pass through P. Every point of a line on the quadric is the vertex of a cone and therefore every line cuts ∞^2 other lines.

The study of the lines, etc., on a quadric is assisted by the process of stereographic projection. If a quadric in S_3 is projected from any point O on the surface on to a plane π not passing through O, the two generators through O are projected into points A, B, and the whole tangent plane at P into the line AB. Any plane section of the quadric is projected into a conic passing through A and B, and any generating line is projected into a line which passes through either A or B. Thus the separation of the generating lines into two systems is made clear.

Taking a V^2_{n-1} and projecting it from a point O on the quadric on to a hyperplane π, the tangent hyperplane τ at O is projected into the S_{n-2} in which it cuts π, and all the lines through O, which form a hypercone V^2_{n-2}, are projected into a quadric V^2_{n-3} in S_{n-2}; the lines, planes, 3-flats, . . . through O are projected into points, lines, planes, . . . of V^2_{n-3}. If S_p is a p-flat of V^2_{n-1}, not passing through O, it cuts τ in an S_{p-1} which also lies in V^2_{n-1}. S_p is projected into S'_p in π, and S_{p-1} into S'_{p-1} which lies in V^2_{n-3}; also S'_p passes through S'_{p-1}. Hence the p-flats in π which correspond to p-flats of V^2_{n-1} pass through $(p-1)$-flats of V^2_{n-3}.

The V^2_{n-3} can then be stereographically projected, and its planes, 3-flats, etc., become the lines, planes, etc., of a V^2_{n-5}, and so on. If n is odd, $= 2p + 1$, we get finally in this way a V^2_0, i.e. a pair of points, and from this we conclude that the lines of the V^2_2 form two separate systems, then the planes of V^2_4 and so on up to the p-flats of V^2_{2p}. If n is even, $= 2p$, we get finally a V_1^2 or conic in which the points form one continuous series, and hence the $(p-1)$-flats of V_{2p-1}^2.

Hence *in* S_{2p+1} *the spaces of maximum dimensions* p *on a quadric form two separate systems, in* S_{2p} *the spaces of maximum dimensions* $p-1$ *form a single continuous system.*

33. A distinction has further to be made in the case where n is odd, $=2p+1$, according as p is odd or even.

In S_3 a V_2^2 has two separate systems of lines, and every line of the one system cuts every line of the other system, but no two lines of the same system intersect.

Consider a V_4^2 in S_5 and let O be any point on the quadric. Project from O on to a hyperplane π. The tangent hyperplane τ at O cuts π in a 3-flat and meets the quadric in a hypercone V_3^2 with vertex O. The lines and planes of V_3^2 which pass through O cut π in the points and lines of a V_2^2. As the lines of V_2^2 form two separate systems so also do the planes of V_4^2 through O; two planes of the same system have only the point O in common; two planes of different systems, through intersecting lines of V_2^2, intersect in a line of V_4^2.

Let α be any plane of V_4^2, not passing through O. α cuts τ in a line l which is projected into a line l' of V_2^2, and α is projected into a plane α' in π, which passes through l'. Hence the planes of π which correspond to planes of V_4^2 are those which pass through lines of V_2^2.

Let α, β be two planes of V_4^2 of the same system. If they have one point in common we can take this point as O, and then we have proved that O is their only common point. Suppose, if possible, that they have no point in common. The corresponding planes α', β' in π cut V_2^2 in non-intersecting lines l', m', and have in general just one point C′ in common. As C′ is not on V_2^2 it corresponds to just one point C on V_4^2 and this point is common to α and β. Hence α, β, two planes of V_4^2 of the same system, have one and only one point in common.

Second, let α, β be two planes of V_4^2 of different systems. If they have one point in common, taking this as O we have seen that they have a line in common. We suppose then that α and β have no point in common so that l and m do not intersect. The corresponding planes α', β' in π cut V_2^2 in intersecting lines and have therefore the point of intersection C′ in common. Since *l* and *m* do not intersect, the line OC′, which is a line of V_4^2, cuts *l* and *m* in different points, and therefore C′ does not correspond to a point of intersection of α and β, and therefore α, β have no point in common. Hence two planes of V_4^2 of the same system either intersect in a line or have no point in common.

We may proceed by induction from V_{2p}^2 to V_{2p+2}^2, assuming the results for V_{2p}^2 which is the quadric in which the lines of V_{2p+2}^2 through a given point O cut the hyperplane π, and the general result may be stated as follows:

A V_{2p}^2 *in* S_{2p+1} *has two different systems of p-flats. If p is even, two p-flats of the same system can intersect only in a space of odd dimensions* (– 1 equivalent to no common point, 1 equivalent to a line in common, and so on), *and two p-flats of different systems can intersect only in a space of even dimensions* (point, plane, etc.). *If p is odd, the reverse is the case.*

REFERENCES

BERTINI, E. Introduzione alla geometria proiettiva degli iperspazi, con appendice sulle curve algebriche e loro singolarità. Pisa, 1907. (2. ed., 1923. German trans. by Duschek. Wien, 1924.)

CAYLEY, A. Chapters in the analytical geometry of (*n*) dimensions. Cambridge Math. J., **4** (1843), 119-127; Math. Papers, i., No. 11.

— On the superlines of a quadric surface in 5-dimensional space. Q.J. Math., **12** (1873), 176-180; Math. Papers, ix., No. 570.

CLIFFORD, W. K. On the classification of loci. Phil. Trans., **169** (1878), 663-681; Math. Papers, No. 33.

CULLIS, C. E. Matrices and determinoids. Cambridge: Univ. Press. Vol. i. (1913), ii. (1918), iii. (part i.) (1925). (Especially vol. ii., p. 78, § 139, and chap. xviii.)

SYLVESTER, J. J. A demonstration of the theorem that every homogeneous quadratic polynomial is reducible by real orthogonal substitutions to the form of a sum of positive and negative squares. Phil. Mag., **4** (1852), pp. 138-142; Math. Papers, i., No. 47. (This paper, which contains the famous Law of Inertia, has no explicit reference to hyperspace; but in certain other of his papers, notably No. 30 of this volume: On certain general properties of homogeneous functions, Camb. and Dublin Math. J., **6** (1851), he makes use of geometrical language; for example he introduces here the term *homaloid* for a flat space. Nos. 25 and 36 and vol. iv., No. 39, also have explicit reference to higher space.)

CHAPTER VI

ANALYTICAL GEOMETRY: METRICAL

1. Metrical Co-ordinates referred to a Fundamental Simplex. The system of projective co-ordinates referred to a fundamental simplex $A_0A_1 \ldots A_n$ can be made metrical in a general way as follows. With the previous notation, $U \equiv (1, 1, \ldots, 1)$ is the unit point, $P \equiv (x_0, x_1, \ldots, x_n)$ an arbitrary point, and UP cuts the hyperplane opposite A_r in L_r. It was proved (Chap. V, § 10) that the cross-ratio

$$(L_0L_r, UP) \equiv X_r = x_r/x_0.$$

Let the distances of U and P from the fundamental hyperplanes be u_r and p_r respectively. Then with the metrical definition of cross-ratio

$$\frac{x_r}{x_0} = \frac{L_0U}{L_0P} \Big/ \frac{L_rU}{L_rP} = \frac{u_0}{p_0} \Big/ \frac{u_r}{p_r}.$$

Hence

$$x_r = \lambda p_r/u_r,$$

where λ is constant. The co-ordinates x_r are therefore certain fixed multiples of the distances of P from the fundamental hyperplanes.

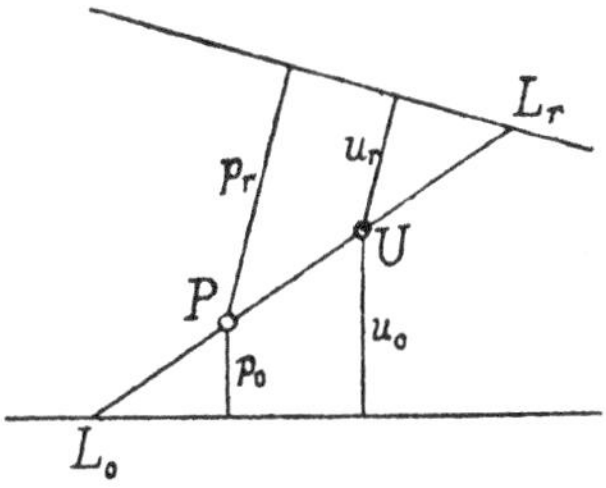

FIG. 11.

Ex. $A_1, A_2, \ldots, A_n$ are n given points in S_n; these are divided in all possible ways into two groups of p and q ($p + q = n$), which determine with two other fixed points, A_0 and U respectively, an S_p and an S_q which intersect in a point T. Prove that the ${}_nC_p$ points T all lie in one hyperplane τ; and that if the hyperplanes (A_1, A_2 . . . A_n) and τ cut the line A_0U in points N and M respectively the cross-ratio $(A_0U, MN) = -p/q$. (S. Kantor, "Torino Atti Acc.," 36 (1901), p. 916.)

[Taking $A_0, A_1, \ldots, A_n$ as fundamental simplex and U as the unit-point the equation of the hyperplane τ is $px_0 - x_1 - x_2 - \ldots - x_n = 0$.]

2. The distances p_r are not independent. If P is joined to the vertices A_r the whole simplex C is divided into $n + 1$ simplexes $S(n + 1)$. If C_r is the content of the simplex $S(n)$ in the fundamental hyperplane opposite A_r, the content of the simplex $S(n + 1)$ with base C_r and vertex P is $\frac{1}{n-1}C_rp_r$ (see Chap. VIII, § 4). Hence

$$\Sigma C_rp_r = (n - 1)C,$$

and

$$\Sigma C_ru_rx_r = \lambda(n - 1)C.$$

3. The equation

$$\Sigma C_ru_rx_r = 0,$$

being homogeneous and linear, represents a certain hyperplane. But this hyperplane does not contain any finite point. The equation is satisfied by ∞^{n-1} sets of finite values of x_r, but these do not correspond to finite values of p_r since λ must here be zero. It therefore represents the *hyperplane at infinity*.

4. Special systems of co-ordinates corresponding to trilinear or areal co-ordinates can be defined, corresponding to a special choice of the point U: viz. for the system corresponding to trilinear co-ordinates we take U equidistant from the fundamental hyperplanes, so that u_r is constant and $x_r = \lambda p_r$; the co-ordinates of a point are proportional to its distances from the fundamental hyperplanes, and the equation of the hyperplane at infinity is $\Sigma C_rx_r = 0$. For the system corresponding to areal co-ordinates we take U such that the lines joining U to the vertices divide the fundamental simplex into simplexes of equal content; then C_ru_r is constant and $x_r = \lambda C_rp_r$; the co-ordinates of a point are proportional to the contents of the simplexes into which the fundamental simplex is divided, and the equation of the hyperplane at infinity is $\Sigma x_r = 0$. The latter is the simplest symmetrical system of metrical co-ordinates, and would be useful in investigating analytically the properties of a simplex.

5. Cartesian Co-ordinates. For metrical geometry, however, the simplest system of metrical co-ordinates is defined by taking the hyperplane at infinity as one of the fundamental hyperplanes, say $x_0 = 0$. We shall then denote the opposite vertex by O and call it the *origin*. The fundamental lines through O, OA_r, are called the co-ordinate-axes, and the fundamental hyperplanes through O the co-ordinate hyperplanes; in each hyperplane there are $n - 1$ axes. As unit point we take a point whose distance from each co-ordinate hyperplane measured along a line parallel to the opposite axis is unity. The co-ordinates $(x_1, x_2, \ldots, x_n)$ of any point are then proportional to its distances from each hyperplane measured along a line parallel to the opposite axis, or, what comes to the same thing, the distance cut off on each axis by a hyperplane parallel to the opposite co-ordinate hyperplane. Then, since the ratio p_0/u_0 becomes equal to unity, the factor of proportionality, λ, becomes simply x_0. The co-ordinates $x_0, x_1, \ldots, x_n$ are then the *general homogeneous cartesian co-ordinates*, with $x_0 = 0$ as hyperplane at infinity. It is usual to deal with non-homogeneous cartesian co-ordinates; these are defined as the ratios x_r/x_0, and are therefore *equal* to the distances cut off on the axes by hyperplanes parallel to the co-ordinate hyperplanes. When the axes OA_r are all mutually orthogonal we have a *rectangular cartesian system.*

We shall consider first the elementary formulæ in non-homogeneous rectangular co-ordinates, which are immediate extensions of those in ordinary geometry.

6. Length of the Radius-vector OP. Let PN be the perpendicular from P on the $(n - 1)$-flat $OA_1 \ldots A_{n-1}$. The co-ordinates of N are $(x_1, x_2, \ldots, x_{n-1}, 0)$. Assuming the formula

$$ON^2 = x_1^2 + x_2^2 + \ldots + x_{n-1}^2$$

to be true for $n - 1$ dimensions, we have, by the theorem of Pythagoras,

$$\rho^2 = OP^2 = ON^2 + NP^2 = x_1^2 + \ldots + x_{n-1}^2 + x_n^2 \quad (6{\cdot}1)$$

Hence by induction this formula is true generally.

The distance, d, between two points $P \equiv (x)$ and $Q \equiv (y)$ is given similarly by

$$d^2 = \Sigma(x_r - y_r)^2 \quad . \quad . \quad . \quad (6\cdot2)$$

7. **Angles.** The angles A_rOP, $= \theta_r$, are called the *direction-angles* of OP, and their cosines are called *direction-cosines*. If OP $= \rho$, we have

$$x_r = \rho \cos \theta_r,$$

and $$\rho^2 = \Sigma x_r^2 = \rho^2 \Sigma \cos^2 \theta_r.$$

Hence $$\Sigma \cos^2 \theta_r = 1 \quad . \quad . \quad . \quad (7\cdot1)$$

Since the projection of OP on any line is equal to the sum of the projections of OA_1, OA_2, . . ., we get, projecting on OP,

$$\rho = \Sigma x_r \cos \theta_r.$$

Let OP′ be any other line and $P' \equiv (x')$; let the direction-angles of OP′ be θ'_r, and the angle POP′ $= \phi$. Then, projecting OP′ on OP, we get

$$\begin{aligned} OP' \cos \phi &= \Sigma x'_r \cos \theta_r \\ &= \Sigma OP' \cos \theta'_r \cos \theta_r. \end{aligned}$$

Hence $$\cos \phi = \Sigma \cos \theta_r \cos \theta'_r \quad . \quad . \quad (7\cdot2)$$

If l_r, l'_r are the direction-cosines of the two lines,

$$\cos \phi = \Sigma l_r l'_r, \qquad \Sigma l_r^2 = 1 = \Sigma l'^2_r.$$

Since

$$\left\| \begin{matrix} l_1 & l_2 & \ldots & l_n \\ l'_1 & l'_2 & \ldots & l'_n \end{matrix} \right\|^2 = \begin{vmatrix} \Sigma l_r^2 & \Sigma l_r l'_r \\ \Sigma l_r l'_r & \Sigma l'^2_r \end{vmatrix} = 1 - (\Sigma l_r l'_r)^2$$

we have

$$\sin^2 \phi = \left\| \begin{matrix} l_1 & \ldots & l_n \\ l'_1 & \ldots & l'_n \end{matrix} \right\|^2 = \Sigma(l_r l'_s - l_s l'_r)^2.$$

Two lines l and l' are orthogonal when

$$\Sigma l_r l'_r = 0,$$

and parallel when

$$l_r = l'_r \quad (r = 1, 2, \ldots, n).$$

If $l_1, \ldots, l_n$ are numbers proportional to the direction-cosines, the actual values of the direction-cosines are found by dividing these by $(\Sigma l_r^2)^{\frac{1}{2}}$, and we have

$$\cos \phi = \frac{\Sigma l_r l'_r}{(\Sigma l_r^2 \, . \, \Sigma l'^2_r)^{\frac{1}{2}}} \quad . \quad . \quad . \quad (7\cdot3)$$

The orientation of an $(n - 1)$-flat is determined by the direction of a normal, and the direction-angles of the $(n - 1)$-flat are those of the normal. The above formulæ then give also the angle between two $(n - 1)$-flats.

8. Equation of an $(n - 1)$-flat. Any equation of the first degree represents an $(n - 1)$-flat. We shall obtain the equation in the canonical form. Through O there is a unique normal ON to the $(n - 1)$-flat, and if P is any point in the $(n - 1)$-flat, NP $\perp$ ON. Let ON $= p$, and the direction-cosines of ON be (l). Let P $\equiv (x)$. Then projecting on OP,

$$\Sigma l_r x_r = p \quad . \quad . \quad . \quad . \quad (8\cdot1)$$

To find the perpendicular distance from a point x' to the hyperplane take x' as origin, writing $x = x' + y$, then the equation of the hyperplane becomes

$$\Sigma l_r(x'_r + y_r) = p \text{ or } \Sigma l_r y_r = p - \Sigma l_r x'_r.$$

The perpendicular from the new origin is the distance p' which we require, hence

$$p' = p - \Sigma l_r x'_r \quad . \quad . \quad . \quad (8\cdot2)$$

9. A straight line is determined by $n - 1$ $(n - 1)$-flats and is represented by $n - 1$ linear equations. It is more conveniently represented, however, by parametric equations. Let c be a fixed point on the line, and let l be its direction-cosines. Then x being an arbitrary point on the line and ρ the distance between the two points, we have

$$x_r = c_r + \rho l_r \quad (r = 1, 2, \ldots, n) \, . \quad . \quad (9\cdot1)$$

Using homogeneous co-ordinates the equations

$$x_r = a_r u + b_r v \quad (r = 0, 1, \ldots, n) \quad . \quad (9\cdot2)$$

are homogeneous parametric equations of the line through the points (a) and (b).

The point at infinity P_∞ on the line is determined by

$$x_0 = 0 = a_0 u + b_0 v.$$

The values of the parameter v/u for the four points a, b, x, P_∞ are 0, ∞, v/u, $-a_0/b_0$. Hence if $a_0 = 1 = b_0$, so that a_r and b_r

are equal to the non-homogeneous co-ordinates of the points a and b, then the cross-ratio

$$(ab, x\mathrm{P}_\infty) = \frac{(ax)}{(bx)} = (0, \infty ; v/u, -1) = -\frac{v}{u}.$$

Hence v/u is equal to the ratio $(ax)/(xb)$, in which x divides the segment ab. The non-homogeneous co-ordinates of the point x are then

$$\frac{x_r}{x_0} = \frac{a_r u + b_r v}{u + v} \quad . \quad . \quad . \quad (9\cdot3)$$

These are Joachimsthal's formulæ for the co-ordinates of a point dividing the join of two points a, b in the ratio v/u.

The equations $x_r = c_r + \rho l_r$ written in homogeneous co-ordinates with homogeneous parameters u, v such that $v/u = \rho$, and with $c_0 = 1$, become

$$\begin{cases} x_r = c_r u + l_r v \quad (r = 1, 2, \ldots, n) \\ x_0 = u. \end{cases}$$

For $v = 0$ we get the point c, for $u = 0$ we get the point at infinity on the line with homogeneous co-ordinates $(0, l_1, l_2, \ldots, l_n)$.

10. The Hypersphere. The locus of a point in S_n which is at a constant distance ρ from a fixed point C is called a *hypersphere of $n - 1$ dimensions.* If the centre $C \equiv (c_1, c_2, \ldots, c_n)$ the equation of the hypersphere is

$$\Sigma(x_r - c_r)^2 = \rho^2 \quad . \quad . \quad . \quad (10\cdot1)$$

This is an equation of the second degree in which the coefficients of the square terms x_r^2 are all unity, and the coefficients of the product terms $x_r x_s$ all zero. Conversely, the equation

$$a\Sigma x_r^2 + 2\Sigma a_r x_r + c = 0 \quad . \quad . \quad (10\cdot2)$$

can be written in the form

$$\Sigma(x_r + a_r/a)^2 = \Sigma a_r^2/a^2 - c/a,$$

and represents a hypersphere with centre $(-a_r/a)$ and radius $(\Sigma a_r^2/a^2 - c/a)^{\frac{1}{2}}$. The homogeneous co-ordinates of the centre are $(-a, a_1, a_2, \ldots, a_n)$ so that if $a = 0$ the centre is a point at infinity, and the radius is infinite. In this case, writing the equation (10·2) in terms of homogeneous co-ordinates

$$a\Sigma x_r^2 + 2\Sigma a_r x_r x_0 + c x_0^2 = 0$$

it reduces to

$$x_0(2\Sigma a_r x_r + cx_0) = 0$$

and represents a finite hyperplane $2\Sigma a_r x_r + cx_0 = 0$ and the hyperplane at infinity $x_0 = 0$.

When $a \neq 0$ we may without loss of generality give it the value unity. Then $\rho^2 = \Sigma a_r^2 - c$. If this is positive the radius is real; if it is negative the radius is imaginary and the hypersphere is said to be *virtual;* if it is zero the hypersphere is called a *point-hypersphere.* In the last case the equation is homogeneous in $(x_r + a_r)$ and represents a hypercone of the first species with vertex $(- a_r)$.

A p-flat in S_n cuts a hypersphere of $n - 1$ dimensions in a hypersphere of $p - 1$ dimensions with real, imaginary, or zero radius.

11. *Every hypersphere of $n - 1$ dimensions cuts the hyperplane at infinity in the same virtual hypersphere of $n - 2$ dimensions.* The equation of the hypersphere in homogeneous co-ordinates being

$$\sum_{r=1}^{n} x_r^2 + 2\sum_{r=1}^{n} a_r x_r x_0 + cx_0^2 = 0,$$

the equations of the section by the hyperplane at infinity $x_0 = 0$ are

$$\sum_{r=1}^{n} x_r^2 = 0, \qquad x_0 = 0.$$

This is the extension to n dimensions of the circular points at infinity in a plane, and the circle at infinity in space, and is called the *hypersphere at infinity.* A plane cuts the hypersphere at infinity in two points, which are the circular points in that plane; and a space of three dimensions cuts it in a circle, which is the circle at infinity for that space. All angular relations can be expressed as projective relations when referred to the hypersphere at infinity. The two lines with direction-cosines l_r and l_r' are orthogonal when $\Sigma l_r l_r' = 0$, but this is the condition that the points at infinity $(0, l_1, \ldots, l_n)$, $(0, l_1', \ldots, l_n')$ on the lines should be conjugate with regard to the hypersphere at infinity. The hypersphere at infinity, together with the hyperplane at infinity counted twice, form a degenerate quadric considered as an envelope and a locus respectively, and is called the *Absolute*.

12. General Cartesian Co-ordinates. In the general system of cartesian co-ordinates the axes may make any angles with one another. Let θ_{rs} be the angle between the axes x_r and x_s, and write $\cos\theta_{rs} = c_{rs}$.

The fundamental formula is the distance-function. In two dimensions we have for the square of the radius vector

$$\rho^2 = \mathrm{OP}^2 = x_1^2 + x_2^2 + 2c_{12}x_1x_2.$$

By induction we can show that in n dimensions

$$\rho^2 = x_1^2 + \ldots + x_n^2 + 2c_{12}x_1x_2 + \ldots + 2c_{n-1,n}x_{n-1}x_n \qquad (12\cdot1)$$

If the line through P parallel to $\mathrm{O}x_r$ cuts the opposite co-ordinate hyperplane in N_r, the co-ordinates of N_r are $(x_1, \ldots, x_{r-1}, 0, x_{r+1}, \ldots, x_n)$. Assuming the formula for ρ^2 to be true for $n - 1$ dimensions we have

$$\mathrm{ON}_n^2 = x_1^2 + \ldots + x_{n-1}^2 + 2c_{12}x_1x_2 + \ldots + 2c_{n-2,n-1}x_{n-2}x_{n-1}.$$

Now

$$\mathrm{OP}^2 = \mathrm{ON}_n^2 + x_n^2 + 2x_n \times (\text{projection of } \mathrm{ON}_n \text{ on the axis of } x_n)$$

and the projection of $\mathrm{ON}_n = \sum_{r=1}^{n-1} c_{rn}x_r$.

Hence

$$\mathrm{OP}^2 = \sum_1^n x_r^2 + 2c_{12}x_1x_2 + \ldots + 2c_{n-2,n-1}x_{n-2}x_{n-1}$$
$$+ 2c_{1n}x_1x_n + \ldots + 2c_{n-1,n}x_{n-1}x_n.$$

Hence the formula is true also for n dimensions.

We can write

$$\rho^2 = \sum_{r=1}^{n}\sum_{s=1}^{n} c_{rs}x_rx_s, \quad . \quad . \quad . \qquad (12\cdot2)$$

since $c_{rs} = 1$ when $r = s$, and $c_{rs} = c_{sr}$.

Similarly the square of the distance between two points (x), (y), is

$$\rho^2 = \Sigma\Sigma c_{rs}(x_r - y_r)(x_s - y_s) \quad . \quad . \qquad (12\cdot3)$$

When ρ and y_r are constant, and x_r variable, these equations represent a hypersphere of radius ρ and centre the origin and the point (y) respectively.

The equations of the absolute, or hypersphere at infinity, are

$$x_0 = 0, \qquad \Sigma\Sigma c_{rs}x_r x_s = 0.$$

13. The direction of the line OP is determined by the ratios of the co-ordinates $x_1, \ldots, x_n$. Denote x_r/ρ by l_r; then l_r are called the *direction-ratios* of OP, so that

$$x_r = \rho l_r \quad . \quad . \quad . \quad . \quad (13{\cdot}1)$$

The direction-ratios are connected by the identical relation

$$\Sigma\Sigma c_{rs}l_r l_s = 1 \quad . \quad . \quad . \quad (13{\cdot}2)$$

If ϕ is the angle between the two lines OP, OP′ with direction-ratios l_r, l'_r, then

$$\begin{aligned} PP'^2 &= \Sigma\Sigma c_{rs}(x_r - x'_r)(x_s - x'_s) \\ &= OP^2 + OP'^2 - 2OP\,.\,OP' \cos\phi. \end{aligned}$$

But $x_r = \rho l_r$ and $x'_r = \rho' l'_r$,
hence

$$\rho^2\Sigma\Sigma c_{rs}l_r l_s + \rho'^2\Sigma\Sigma c_{rs}l'_r l'_s - 2\rho\rho'\Sigma\Sigma c_{rs}l_r l'_s = \rho^2 + \rho'^2 - 2\rho\rho' \cos\phi\,;$$

therefore

$$\cos\phi = \Sigma\Sigma c_{rs}l_r l'_s \quad . \quad . \quad . \quad (13{\cdot}3)$$

The condition that the two lines be orthogonal is

$$\Sigma\Sigma c_{rs}l_r l'_s = 0,$$

which is the condition that the two points at infinity l_r, l'_r should be conjugate with regard to the absolute.

As only the ratios of l_r are required, an expression for $\cos\phi$ in terms of these ratios is preferable. This is

$$\cos\phi = \frac{\Sigma\Sigma c_{rs}l_r l'_s}{(\Sigma\Sigma c_{rs}l_r l_s\,.\,\Sigma\Sigma c_{rs}l'_r l'_s)^{\frac{1}{2}}} \quad . \quad . \quad (13{\cdot}4)$$

If ϕ_r is the angle which the line (l) makes with the axis of x_r, we have from (13·4), putting $l'_s = 0$, except when $s = r$,

$$\cos\phi_r = \Sigma c_{rs}l_s/(\Sigma\Sigma c_{ij}l_i l_j)^{\frac{1}{2}} \quad . \quad . \quad (13{\cdot}5)$$

14. Equation of a Hyperplane. Let ON ($= p$) be the normal from O to the hyperplane, and let the direction-ratios of ON be l_r, and let $P \equiv (x)$ be any point on the hyperplane. Projecting the co-ordinates of P on to ON we get

$$p = \Sigma x_r \cos \phi_r = \Sigma\Sigma c_{rs} l_s x_r / (\Sigma\Sigma c_{rs} l_r l_s)^{\frac{1}{2}}, \qquad (14{\cdot}1)$$

an equation of the first degree. If the equation of the hyperplane is

$$\Sigma \xi_r x_r = k, \quad . \quad . \quad . \qquad (14{\cdot}2)$$

the direction-ratios l_r of its normal and the length of the perpendicular p from the origin are given by the equations

$$\lambda = \frac{\Sigma c_{rs} l_s}{\xi_r} = \frac{p(\Sigma\Sigma c_{ij} l_i l_j)^{\frac{1}{2}}}{k} \quad (r = 1, 2, \ldots, n)$$

Denote the determinant $| c_{nn} |$ by C, and the co-factor of c_{rs} by C_{rs}. Then

$$\lambda \sum_{r=1}^{n} C_{ir} \xi_r = C l_i, \quad . \quad . \quad . \qquad (14{\cdot}3)$$

which determines the direction-ratios l_i.

Then

$$C \sum_{r=1}^{n} \sum_{s=1}^{n} c_{rs} l_s l_r = C\lambda \sum_{r=1}^{n} l_r \xi_r = \lambda^2 \sum_{i=1}^{n} \sum_{j=1}^{n} C_{ij} \xi_j \xi_i,$$

and
$$C \sum_{r=1}^{n} \sum_{s=1}^{n} c_{rs} l_r l'_s = \lambda^2 \sum_{i=1}^{n} \sum_{j=1}^{n} C_{ij} \xi_i \xi'_j.$$

Hence
$$p = \frac{kC^{\frac{1}{2}}}{(\Sigma\Sigma C_{ij} \xi_i \xi_j)^{\frac{1}{2}}} \quad . \quad . \quad . \qquad (14{\cdot}4)$$

15. The angle ϕ between two hyperplanes ξ, ξ' is equal to the angle between their normals, hence

$$\cos \phi = \frac{\Sigma\Sigma C_{rs} \xi_r \xi'_s}{(\Sigma\Sigma C_{rs} \xi_r \xi_s \,.\, \Sigma\Sigma C_{rs} \xi'_r \xi'_s)^{\frac{1}{2}}},$$

i.e. the expression for the angle between two hyperplanes is the same as that for two lines with C_{rs} instead of c_{rs}; and the condition that the two hyperplanes should be orthogonal is

$$\Sigma\Sigma C_{rs} \xi_r \xi'_s = 0.$$

This is the condition that they should be conjugate with regard to the quadric whose tangential equation is

$$\Sigma\Sigma C_{rs}\xi_r^2 = 0.$$

Now this is just the tangential equation of the absolute, or the condition that the hyperplane $\Sigma\xi_r x_r = 0$ should touch the cone $\Sigma\Sigma c_{rs}x_r x_s = 0$.

16. Plücker Co-ordinates. We shall consider now in greater detail the analytical metrical geometry in space of four dimensions.

A plane through the origin is represented by two equations

$$\left.\begin{aligned} l_1x_1 + l_2x_2 + l_3x_3 + l_4x_4 &= 0,\\ m_1x_1 + m_2x_2 + m_3x_3 + m_4x_4 &= 0. \end{aligned}\right\} \quad . \quad (16{\cdot}1)$$

These involve six constants, but a plane through a point has just four degrees of freedom. The same plane is determined if we take instead of l_1, l_2, l_3, l_4 the numbers $l_1 + \lambda m_1$, $l_2 + \lambda m_2$, etc., where λ is any parameter. A symmetrical set of co-ordinates for the plane is got by taking the six expressions

$$l_i m_j - l_j m_i = p_{ij} \quad (i, j = 1, 2, 3, 4) \quad . \quad (16{\cdot}2)$$

where $p_{ij} = -p_{ji}$. The ratios of these six quantities are the same whatever pair of hyperplanes we take to fix the same plane, for

$$(l_i + \lambda m_i)(l_j + \mu m_j) - (l_j + \lambda m_j)(l_i + \mu m_i) = (\mu - \lambda)(l_i m_j - l_j m_i).$$

Eliminating x_1, x_2, x_3 and x_4 in succession between the two equations (16·1) we get

$$\left.\begin{aligned} p_{12}x_2 + p_{13}x_3 + p_{14}x_4 &= 0,\\ p_{21}x_1 \qquad\quad + p_{23}x_3 + p_{24}x_4 &= 0,\\ p_{31}x_1 + p_{32}x_2 \qquad\quad + p_{34}x_4 &= 0,\\ p_{41}x_1 + p_{42}x_2 + p_{43}x_3 \qquad\quad &= 0. \end{aligned}\right\} \quad . \quad . \quad (16{\cdot}3)$$

These represent four hyperplanes all passing through the same plane. Eliminating x_1 and x_2 from the first three equations of (16·3) by multiplying them respectively by p_{23}, p_{31}, p_{12} and adding, x_3 also disappears, and we obtain the relation

$$p_{23}p_{14} + p_{31}p_{24} + p_{12}p_{34} = 0 \quad . \quad . \quad (16{\cdot}4)$$

Hence by this identity the six quantities p are equivalent to only four independent co-ordinates. Further, any set of six quantities connected by the identical relation (16·4) uniquely determine a plane, for this is represented by any two of the equations (16·3).

The plane can also be represented by parametric equations

$$x_i = ua_i + vb_i \quad (i = 1, 2, 3, 4). \quad . \quad (16·5)$$

where a and b are two points which, with the origin, determine the plane. For a and b can be substituted any point on the lines Oa and Ob respectively, so that only the ratios of the co-ordinates of the two points are required, hence we have again six constants. Again for a we may substitute any point on the line ab, i.e. for a_1, a_2, a_3, a_4 we may substitute $a_1 + \lambda b_1$, etc., where λ is any parameter. A symmetrical set of co-ordinates for the plane is got by taking the six expressions

$$a_i b_j - a_j b_i = \varpi_{ij} \quad (i, j = 1, 2, 3, 4) \quad . \quad (16·6)$$

and the ratios of these six quantities are the same whatever pair of points we take to fix the same plane through O.

Since each of the two points lies in each of the two hyperplanes we have

$$\Sigma l_r a_r = 0, \quad \Sigma l_r b_r = 0, \quad \Sigma m_r a_r = 0, \quad \Sigma m_r b_r = 0.$$

Eliminating l_4 from the first two, and m_4 from the last two, we get

$$l_1\varpi_{14} + l_2\varpi_{24} + l_3\varpi_{34} = 0,$$

and

$$m_1\varpi_{14} + m_2\varpi_{24} + m_3\varpi_{34} = 0;$$

hence the ratios of ϖ_{14}, ϖ_{24}, ϖ_{34} are determined, viz. :—

$$\varpi_{14} : \varpi_{24} : \varpi_{34} = p_{23} : p_{31} : p_{12}.$$

Similarly we can prove that

$$\varpi_{14} : \varpi_{24} : \varpi_{34} : \varpi_{23} : \varpi_{31} : \varpi_{12} = p_{23} : p_{31} : p_{12} : p_{14} : p_{24} : p_{34}.$$

The two sets of co-ordinates are therefore equivalent.

17. Condition that two Planes p, p' through the Origin should have a Line in Common. The condition is that the four hyperplanes l, m, l', m' should have a line in common, i.e.

$$\begin{vmatrix} l_1 & l_2 & l_3 & l_4 \\ m_1 & m_2 & m_3 & m_4 \\ l'_1 & l'_2 & l'_3 & l'_4 \\ m'_1 & m'_2 & m'_3 & m'_4 \end{vmatrix} = 0.$$

When this is expanded it gives

$$p_{14}p'_{23} + p_{24}p'_{31} + p_{34}p'_{12} + p_{23}p'_{14} + p_{31}p'_{24} + p_{12}p'_{34} = 0.$$

18. Representation by Points on a Quadric in Five Dimensions. By making the two planes coincide, we obtain again the identity

$$\phi \equiv p_{14}p_{23} + p_{24}p_{31} + p_{34}p_{12} = 0.$$

The co-ordinates p or ϖ are precisely the same as Plücker's co-ordinates of a line in S_3. In fact the geometry of planes through a point in S_4 is projectively the same as that of lines in S_3. If we take p_{ij} as the homogeneous co-ordinates of a point in S_5, the identical relation ϕ, which is of the second degree, represents a quadric. Thus lines in S_3, and planes through a point in S_4, can be represented by points on a certain quadric in S_5. If we make the real transformation

$$p_{14} = y_1 + y_4, \quad p_{24} = y_2 + y_5, \quad p_{34} = y_3 + y_6,$$
$$p_{23} = y_1 - y_4, \quad p_{31} = y_2 - y_5, \quad p_{12} = y_3 - y_6,$$

ϕ becomes

$$y_1^2 + y_2^2 + y_3^2 - y_4^2 - y_5^2 - y_6^2 = 0,$$

and therefore the quadric ϕ in S_5 contains two systems of real planes. (Chap. v, § 22.)

The condition that two planes in S_4 should intersect in a line is that their corresponding points on ϕ should be conjugate, and hence their join lies entirely in ϕ. Hence a line in ϕ represents a singly infinite system of planes in S_4 all passing through the same line, i.e. a pencil of planes; hence also it represents a plane pencil of lines in S_3. A plane in ϕ similarly represents a doubly infinite system of lines in S_3 of which every two intersect, i.e. either a system of lines all passing through the same point (bundle of lines), or a system of lines all in one plane (plane field of lines); and similarly it represents in S_4 either a sheaf of planes all passing through the same line, or a system of planes all lying in the same 3-flat and of course passing through O, i.e. a bundle of planes in space of three dimensions. These two different systems, in S_3 or in S_4, correspond to the two separate systems of planes in ϕ.

Two bundles of lines in S_3 have just one line in common, that which joins the vertices; two plane fields of lines in S_3 have also just one line in common, the line of intersection of their planes; in S_4 two sheaves of planes through intersecting lines have just one plane in common, that determined by their axes; and two 3-dimensional bundles of planes in S_4 through a common point have just one plane in common, the plane common to the two 3-flats:—these properties correspond to the fact that two planes of ϕ of the same system have one and only one point in common.

Again, a bundle of lines, and a plane field of lines, in S_3, have either a plane pencil of lines in common when the vertex of the bundle lies in the plane of the field, or have no line in common; a sheaf of planes through a line in S_4, and a bundle of planes in a 3-flat of S_4, whose vertex lies in the axis of the sheaf, have either a pencil of planes in common when the axis of the sheaf lies in the 3-flat of the bundle, or have no plane in common: these properties correspond to the fact that two planes of ϕ of different systems have either no point in common or intersect in a line.

19. All the planes in S_4 through O which cut a given plane through O in straight lines form a 3-dimensional assemblage represented by the section of ϕ by a tangent hyperplane; this is a hypercone of the first species whose vertex (the point of contact) corresponds to the fixed plane. In S_3 the corresponding system is the system of lines all cutting a given line.

In S_3 a system of lines corresponding to a single homogeneous equation in p_{ij}, and therefore depending on three parameters, is called a *complex* of lines; two equations in p_{ij} represent a *congruence;* and three equations a *line-series.* When the equations are linear we have the *linear complex*, *linear congruence*, and *regulus.* A linear complex is represented by a hyperplane section of ϕ, i.e. by a V_3^2 on ϕ. When the hyperplane is tangent to ϕ the complex is called a *special* linear complex. A linear congruence is represented by a V_2^2 on ϕ, and a linear series by a conic. Since ϕ is of the second class as well as second degree, through any 3-flat there pass two tangent 4-flats. Hence the 3-flat of any V_2^2 on ϕ is the intersection of two tangent 4-flats, and therefore a linear congruence is in general compounded of two special linear complexes, i.e. it consists of lines which intersect two fixed lines; these are called its *directrices*, and they correspond to the points of contact of the two 4-flats. Similarly a linear series consists of lines which intersect three fixed lines, and is therefore a regulus or system of rectilinear generators of a quadric surface.

20. A special type of linear congruence corresponds to sections of ϕ by tangent 3-flats. If the 3-flat is tangent in a line it meets the quadric in a hypercone having this line double, i.e. two planes through the line. The directrices of the congruence are represented by any two points on the line, and as this line lies in ϕ the two points are conjugate and therefore the directrices intersect. The congruence thus consists of all lines which cut two intersecting lines; if A is the point and α the plane common to the two directrices, the two planes of ϕ which represent the congruence correspond to the bundle of lines through A and the plane field of lines in α.

If the 3-flat is tangent in a point it meets ϕ in a cone. In the congruence the two directrices coincide and form one line a, which corresponds to the vertex of the cone, and cuts every other line of the congruence; the other lines form a system of plane pencils, a being a line of each pencil, and therefore the planes of the pencils all pass through a. This congruence is called the *special* linear congruence.

21. Metrical Relations between Planes through a Point in S_4. Metrical relations are relations referred to the absolute. Taking rectangular cartesian co-ordinates, the point-equations of the absolute are

$$x_1^2 + x_2^2 + x_3^2 + x_4^2 = 0, \quad x_0^2 = 0.$$

If a and b are two points on a plane through O, the absolute polars of a and b are

$$a_1x_1 + a_2x_2 + a_3x_3 + a_4x_4 = 0,$$
$$b_1x_1 + b_2x_2 + b_3x_3 + b_4x_4 = 0,$$

and the intersection of these two hyperplanes is the absolute polar of the given plane. The Plücker co-ordinates of the absolute polar of the plane p are therefore

$$p'_{23} = a_2b_3 - a_3b_2 = \varpi_{23} = \lambda p_{14}, \text{ etc.},$$

i.e. the absolute polar of the plane

$$(p_{23}, p_{31}, p_{12}, p_{14}, p_{24}, p_{34})$$

is

$$(p_{14}, p_{24}, p_{34}, p_{23}, p_{31}, p_{12}).$$

22. Condition that two Planes should have a Line in Common. Let the planes be determined by the pairs of hyperplanes (summations being from 0 to 4)

$$\left.\begin{aligned}\Sigma l_r x_r &= 0,\\ \Sigma m_r x_r &= 0,\end{aligned}\right\} \qquad \left.\begin{aligned}\Sigma l'_r x_r &= 0,\\ \Sigma m'_r x_r &= 0.\end{aligned}\right\}$$

Then it must be possible to determine λ and λ' so that

$$\Sigma l_r x_r + \lambda \Sigma m_r x_r = 0,$$

and

$$\Sigma l'_r x_r + \lambda' \Sigma m'_r x_r = 0$$

should coincide. Therefore

$$\frac{l_0 + \lambda m_0}{l'_0 + \lambda' m'_0} = \cdot \cdot \cdot = \frac{l_4 + \lambda m_4}{l'_4 + \lambda' m'_4} = k,$$

i.e. $l_r + \lambda m_r - k l'_r - k\lambda' m'_r = 0 \quad (r = 0, \ldots, 4)$.

Eliminating λ, k, $k\lambda'$ we have

$$\left\| \begin{matrix} l_0 & l_1 & \ldots & l_4 \\ m_0 & m_1 & \ldots & m_4 \\ l'_0 & l'_1 & \ldots & l'_4 \\ m'_0 & m'_1 & \ldots & m'_4 \end{matrix} \right\|_4 = 0,$$

i.e. this matrix is of rank 3. This is equivalent to 2 conditions.

23. Conditions for Parallelism. Two planes are *completely parallel* when they intersect the hyperplane at infinity in the same line, hence the 5 hyperplanes

$$\Sigma l_r x_r = 0, \quad \Sigma m_r x_r = 0, \quad \Sigma l'_r x_r = 0, \quad \Sigma m'_r x_r = 0, \quad x_0 = 0$$

have a line in common. The condition for this is that the matrix

$$\begin{bmatrix} l_0 & l_1 & \ldots & l_4 \\ m_0 & m_1 & \ldots & m_4 \\ l'_0 & l'_1 & \ldots & l'_4 \\ m'_0 & m'_1 & \ldots & m'_4 \\ 1 & 0 & \ldots & 0 \end{bmatrix}$$

should be of rank 3, and therefore the matrix

$$\begin{bmatrix} l_1 & l_2 & l_3 & l_4 \\ m_1 & m_2 & m_3 & m_4 \\ l'_1 & l'_2 & l'_3 & l'_4 \\ m'_1 & m'_2 & m'_3 & m'_4 \end{bmatrix}$$

should be of rank 2. This is equivalent to 4 conditions.

24. Two planes are *half-parallel* when they intersect in one point at infinity and have no finite point in common. The condition for this is that

$$\begin{vmatrix} l_1 & l_2 & l_3 & l_4 \\ m_1 & m_2 & m_3 & m_4 \\ l'_1 & l'_2 & l'_3 & l'_4 \\ m'_1 & m'_2 & m'_3 & m'_4 \end{vmatrix} = 0,$$

i.e. that the matrix should be of rank 3 (one condition).

In terms of the Plücker co-ordinates this is equivalent to

$$p_{23}p'_{14} + p_{31}p'_{24} + p_{12}p'_{34} + p_{14}p'_{23} + p_{24}p'_{31} + p_{34}p'_{12} = 0,$$

which expresses that the lines at infinity on the two planes intersect. If the planes both pass through O this is the condition that the planes should intersect in a line.

25. Conditions for Orthogonality. Two planes are *completely orthogonal* when the line at infinity on each coincides with the absolute polar of the line at infinity on the other. In terms of the Plücker co-ordinates this is expressed by the equations

$$\frac{p_{23}}{p'_{14}} = \frac{p_{31}}{p'_{24}} = \frac{p_{12}}{p'_{34}} = \frac{p_{14}}{p'_{23}} = \frac{p_{24}}{p'_{31}} = \frac{p_{34}}{p'_{12}}.$$

This is equivalent to 4 conditions, for if five of these ratios are equated the equality of the sixth follows from the identical relation between the co-ordinates.

26. Two planes are *half-orthogonal* when the line at infinity on each intersects the absolute polar of the line at infinity on the other. The condition for this is

$$p_{23}p'_{23} + p_{31}p'_{31} + p_{12}p'_{12} + p_{14}p'_{14} + p_{24}p'_{24} + p_{34}p'_{34} = 0$$

(one condition).

Ex. Show that the two planes

$$\left.\begin{array}{l} x_3 = 0, \\ x_4 = 0, \end{array}\right\} \quad \text{and} \quad \left.\begin{array}{l} x_0 + m_1x_1 + m_2x_2 = 0, \\ x_0 + m_3x_3 + m_4x_4 = 0 \end{array}\right\}$$

are both half-parallel and half-orthogonal.

27. Consider two planes p and p' through O, and let q and q' be their absolute polars. Then there are two planes through

O which cut these four planes in straight lines. If these planes are u, v we have, to determine u or v, the equations

$$\begin{aligned} p_{23}u_{14} + p_{31}u_{24} + p_{12}u_{34} + p_{14}u_{23} + p_{24}u_{31} + p_{34}u_{12} &= 0, \\ p'_{23}u_{14} + \dots &= 0, \\ p_{14}u_{14} + p_{24}u_{24} + p_{34}u_{34} + p_{23}u_{23} + p_{31}u_{31} + p_{12}u_{12} &= 0, \\ p'_{14}u_{14} + \dots &= 0, \end{aligned}$$

and also

$$u_{23}u_{14} + u_{31}u_{24} + u_{12}u_{34} = 0.$$

These five equations, of which four are linear and one quadratic, determine two sets of values of the ratios u_{ij}.

If the two planes are isocline, they have an infinity of common perpendicular planes. The condition for this is that the matrix

$$\begin{bmatrix} p_{23} & p_{31} & p_{12} & p_{14} & p_{24} & p_{34} \\ p'_{23} & p'_{31} & p'_{12} & p'_{14} & p'_{24} & p'_{34} \\ p_{14} & p_{24} & p_{34} & p_{23} & p_{31} & p_{12} \\ p'_{14} & p'_{24} & p'_{34} & p'_{23} & p'_{31} & p'_{12} \end{bmatrix}$$

should be of rank 3. This is equivalent to two single conditions.

Ex. Show that the plane $(a, b, c;\ a, b, c')$ is isocline to the plane $x_3 = 0 = x_4$, and prove that if ϕ is the angle between them, $\tan^2 \phi = -c/c'$.

28. When the Plücker co-ordinates p_{ij} are given, the coefficients in the equations of the two hyperplanes which determine the plane still admit of a variety of values. If $p_{12} \neq 0$ we may take $l_1 = 0$, $l_2 = 1$, $m_1 = 1$, $m_2 = 0$, then $l_3 = p_{31}$, $l_4 = -p_{14}$, $m_3 = p_{23}$, $m_4 = p_{24}$.

The problem to calculate the angles between two planes can be illustrated with a numerical example.

Let the given planes be

$$p\begin{cases} x_1 + 7x_2 + x_3 = 0, \\ x_4 = 0, \end{cases} \qquad p'\begin{cases} x_1 + x_2 = 0, \\ x_3 + x_4 = 0, \end{cases}$$

and let q, q' be their absolute polars.

Then the Plücker co-ordinates are

	23	31	12	14	24	34
p	0	0	0	1	7	1
p'	1	−1	0	1	1	0
q	1	7	1	0	0	0
q'	1	1	0	1	−1	0

Let u_{ij} be the co-ordinates of a plane cutting these four planes in lines. Then

$$\begin{aligned} u_{14} + 7u_{24} + u_{34} &= 0 \\ u_{23} - u_{31} + u_{14} + u_{24} &= 0 \\ u_{23} + 7u_{31} + u_{12} &= 0 \\ u_{23} + u_{31} + u_{14} - u_{24} &= 0 \end{aligned}$$

Hence $u_{14} = -u_{23}$, $u_{24} = u_{31}$, $u_{34} = u_{23} - 7u_{31}$, $u_{12} = -u_{23} - 7u_{31}$. Substituting in the equation

$$u_{23}u_{14} + u_{31}u_{24} + u_{12}u_{34} = 0$$

in terms of u_{23} and u_{31}, we have

$$-u_{23}^2 + u_{31}^2 - u_{23}^2 + 49u_{31}^2 = 0,$$

i.e. $$u_{23}^2 = 25u_{31}^2, \quad \text{and} \quad u_{23} = \pm 5u_{31}.$$

Hence the two common orthogonal planes are

	23	31	12	14	24	34
u	5	1	− 12	− 5	1	− 2
v	− 5	1	− 2	5	1	− 12

We can express these as

$$u\begin{cases} 12x_2 + x_3 + 5x_4 = 0, \\ x_1 - 5x_2 - 2x_4 = 0, \end{cases} \quad v\begin{cases} 2x_2 + x_3 - 5x_4 = 0, \\ x_1 + 5x_2 - 12x_4 = 0. \end{cases}$$

We have next to find the direction-cosines of the lines of intersection of p and p' with both u and v. For the line of intersection of p and u we take any three of the four equations in x_1, x_2, x_3, x_4 whose coefficients are

$$\begin{matrix} 1, & 7, & 1, & 0, \\ 0, & 0, & 0, & 1, \\ 0, & 12, & 1, & 5, \\ 1, & -5, & 0, & -2. \end{matrix}$$

Hence we find

$$(pu) \equiv (\ 5, \ 1, \ -12, \ 0).$$

Similarly

$$\begin{aligned} (p'u) &\equiv (\ 1, \ -1, \ -3, \ 3), \\ (pv) &\equiv (-5, \ 1, \ -2, \ 0), \\ (p'v) &\equiv (\ 3, \ -3, \ 1, \ -1). \end{aligned}$$

The two angles θ and ϕ between the planes are the angles between (pu) and $(p'u)$, (pv) and $(p'v)$. Hence

$$\cos\theta = \frac{40}{\sqrt{(170 \times 20)}} = \frac{4}{\sqrt{34}}, \quad \cos\phi = \frac{-20}{\sqrt{(30 \times 20)}} = -\frac{2}{\sqrt{6}}.$$

Ex. Show that the two planes $\left.\begin{aligned} x_1 + x_2 + x_3 = 0, \\ x_4 = 0, \end{aligned}\right\}$ and $\left.\begin{aligned} x_1 + x_2 = 0, \\ x_3 + x_4 = 0, \end{aligned}\right\}$ intersect in a line, and find their dihedral angle. $\left(\text{Ans. } \cos^{-1}\frac{1}{\sqrt{3}}.\right)$

29. Co-ordinates of a $(k - 1)$-flat. A $(k - 1)$-flat in S_n can be determined by co-ordinates analogous to Plücker's co-ordinates of a straight line in S_3. These were first introduced by Grassmann, and will be called Grassmann's co-ordinates. The $(k - 1)$-flat is determined by k points with co-ordinates

$$\begin{array}{cccc} (a_{00}, & a_{01}, & \ldots, & a_{0n}), \\ \cdot & \cdot & \cdot & \cdot \\ (a_{k-1,0}, & a_{k-1,1}, & \ldots, & a_{k-1,n}), \end{array}$$

and we may take as co-ordinates of the $(k - 1)$-flat the k-rowed determinants of this matrix, the number of which is ${}_{n+1}C_k$. Denote the determinant obtained by choosing the columns numbered $\alpha, \beta, \gamma, \ldots$ (i.e. having these numbers as the second suffix) by $p_{\alpha\beta\gamma\ldots}$, the suffixes $\alpha, \beta, \gamma, \ldots$ being arranged in ascending order of magnitude.

The $(k - 1)$-flat is also determined by $n - k + 1$ equations

$$\begin{array}{lllll} l_{00}x_0 & + l_{01}x_1 & + \ldots & + l_{0n}x_n & = 0, \\ \cdot & \cdot & \cdot & \cdot & \cdot \\ l_{n-k,0}x_0 & + l_{n-k,1}x_1 & + \ldots & + l_{n-k,n}x_n & = 0, \end{array}$$

and we may take as co-ordinates of the $(k - 1)$-flat the $(n - k + 1)$-rowed determinants of the matrix of these equations, the number of which is also ${}_{n+1}C_k$. Denote the determinant obtained by choosing the columns numbered $\alpha, \beta, \gamma, \ldots$ by $\varpi_{\alpha\beta\gamma\ldots}$

We can show that these two sets of co-ordinates are proportional. Since each of the k points lies on each of the $n - k + 1$ hyperplanes we have

$$\begin{array}{lllll} l_{00}a_{00} & + l_{01}a_{01} & + \ldots & + l_{0n}a_{0n} & = 0, \\ \cdot & \cdot & \cdot & \cdot & \cdot \\ l_{n-k,0}a_{00} & + l_{n-k,1}a_{01} & + \ldots & + l_{n-k,n}a_{0n} & = 0. \end{array}$$

Multiply these equations respectively by the co-factors of $b_0, b_1, \ldots, b_{n-k}$ in the determinant

$$\begin{vmatrix} b_0 & l_{0,k+1} & \ldots & l_{0n} \\ \cdot & \cdot & \cdot & \cdot \\ b_{n-k} & l_{n-k,k+1} & \ldots & l_{n-k,n} \end{vmatrix}$$

and add. The last $n - k$ terms vanish and we get

$$a_{00}p_{0u} + a_{01}p_{1u} + \ldots + a_{0k}p_{ku} = 0,$$

where u stands for the succession of suffixes $k + 1, \ldots, n$.

Similarly

$$a_{10}p_{0u} + a_{11}p_{1u} + \ldots + a_{1k}p_{ku} = 0,$$
$$\ldots \quad \ldots \quad \ldots$$
$$a_{k-1,0}p_{0u} + a_{k-1,1}p_{1u} + \ldots + a_{k-1,k}p_{ku} = 0.$$

These k equations determine the ratios of $p_{0u}, \ldots, p_{ku}$, and we have

$$p_{0,k+1,\ldots,n} = \lambda\varpi_{1,2,\ldots,k}, \text{ etc.}$$

and generally

$$p_{abc\ldots} = \lambda\varpi_{\alpha\beta\gamma\ldots},$$

where $abc \ldots \alpha\beta\gamma \ldots$ is an even permutation of the numbers $0, 1, 2, \ldots, n$; i.e. each co-ordinate of the one set is proportional to the co-ordinate of the other set with the complementary suffixes.

30. A k-flat in S_n has $(k+1)(n-k)$ degrees of freedom. We have ${}_{n+1}C_{k+1}$ homogeneous co-ordinates. These are connected by ${}_{n+1}C_{k+1} - 1 - (k+1)(n-k)$ independent identities.

Consider the matrix

$$\begin{bmatrix} a_{00} & a_{01} & \ldots & a_{0,n-k+1} & a_{0,n-k} & \ldots & a_{0n} \\ \cdot & \cdot & \cdot & \cdot & \cdot & \cdot & \cdot \\ a_{k0} & a_{k1} & \ldots & a_{k,n-k-1} & a_{k,n-k} & \ldots & a_{kn} \end{bmatrix}$$

and denote by A_r the co-factor of a_r in the determinant

$$\begin{vmatrix} a_0 & a_{0,n-k+1} & \ldots & a_{0n} \\ \cdot & \cdot & \cdot & \cdot \\ a_k & a_{k,n-k+1} & \ldots & a_{kn} \end{vmatrix}.$$

Then if u denotes the succession of suffixes $n-k+1, \ldots, n$,

$$p_{0u} = a_{00}A_0 + a_{10}A_1 + \ldots + a_{k0}A_k,$$
$$-p_{1u} = a_{01}A_0 + a_{11}A_1 + \ldots + a_{k1}A_k,$$
$$\ldots \quad \ldots \quad \ldots$$
$$(-)^{n-k}p_{n-k,u} = a_{0,n-k}A_0 + a_{1,n-k}A_1 + \ldots + a_{k,n-k}A_k,$$
$$0 = a_{0,n-k+1}A_0 + a_{1,n-k+1}A_1 + \ldots + a_{k,n-k+1}A_k,$$
$$\ldots \quad \ldots \quad \ldots$$
$$0 = a_{0n}A_0 + a_{1n}A_1 + \ldots + a_{kn}A_k.$$

If we take the last k equations with any two of the others and eliminate the A's we get a relation which is satisfied identically, e.g. taking the first two equations with the last k and eliminating we get merely $p_{0u}p_{1u} - p_{1u}p_{0u} = 0$. But if we take three of the first set with $k - 1$ of the last, e.g. the first three with the last $k - 1$, we get a three-termed relation

$$p_{0, n-k+1, \ldots, n}\, p_{1, 2, n-k+2, \ldots, n} - p_{1, n-k+1, \ldots, n}\, p_{0, 2, n-k+2, \ldots, n}, \\ - p_{2, n-k+1, \ldots, n}\, p_{0, 1, n-k+2, \ldots, n} = 0.$$

These relations are all of the type

$$p_{adi \ldots j}\; p_{bci \ldots j} + p_{bdi \ldots j}\; p_{cai \ldots j} + p_{cdi \ldots j}\; p_{abi \ldots j} = 0,$$

where $i \ldots j$ denotes a permutation of $k - 1$ of the numbers $0, 1, \ldots, n$ and a, b, c, d are four of the remaining ones.

31. A p-flat and a q-flat in S_{p+q+1} in general do not intersect, but require one condition for intersection in a point.

The p-flat is represented by $q + 1$ equations

$$a_{00}x_0 + a_{01}x_1 + \ldots + a_{0, p+q+1}x_{p+q+1} = 0,$$
$$\cdot \quad \cdot \quad \cdot \quad \cdot \quad \cdot \quad \cdot \quad \cdot \quad \cdot \quad \cdot$$
$$a_{q0}x_0 + a_{q1}x_1 + \ldots + a_{q, p+q+1}x_{p+q+1} = 0,$$

and the q-flat by $p + 1$ equations

$$b_{00}x_0 + b_{01}x_1 + \ldots + b_{0, p+q+1}x_{p+q+1} = 0,$$
$$\cdot \quad \cdot \quad \cdot \quad \cdot \quad \cdot \quad \cdot \quad \cdot \quad \cdot \quad \cdot$$
$$b_{p0}x_0 + b_{p1}x_1 + \ldots + b_{p, p+q+1}x_{p+q+1} = 0.$$

Eliminating $x_0, \ldots, x_{p+q+1}$ between these $p + q + 2$ equations we get the result in the form of a determinant which when expanded gives the condition

$$\Sigma p_{ijk} \ldots q_{lmn} \ldots = 0,$$

where $ijk \ldots lmn \ldots$ is an even permutation of the suffixes $0, 1, \ldots, p + q + 1$, the first set containing $q + 1$ numbers and the other set $p + 1$, and the summation extends to all the different partitions of the $p + q + 2$ numbers into two such sets. The condition for intersection is thus linear and homogeneous in the co-ordinates of the p-flat and the q-flat.

32. In S_n a k-flat has ${}_{n+1}C_{k+1}$ homogeneous co-ordinates, and if these are taken as point-co-ordinates in a space of $N = {}_{n+1}C_{k+1} - 1$ dimensions the assemblage of k-flats is represented by points on a variety V_M of $M = (k + 1)(n - k)$

dimensions in this S_N. V_M is cut by an S_{N-M} in points, the number R of such points being the order of the variety. Now an S_{N-M} is represented by M linear equations in the co-ordinates, and each such equation represents the condition that the k-flat in S_n should intersect an $(n-k-1)$-flat. Hence the order R of V_M^R is equal to the number of k-flats in S_n which cut $(k+1)(n-k)$ arbitrary $(n-k-1)$-flats.

It has been proved by Schubert (see Chap. I, § 24) that the number

$$R = \frac{1!\,2!\,3!\,\ldots\,k!\,\{(k+1)(n-k)\}!}{n!\,(n-1)!\,(n-2)!\,\ldots\,(n-k)!}.$$

REFERENCES

BAKER, H. F. Principles of geometry. Cambridge: Univ. Press. Vol. i. Foundations (1922); Vol. ii. Plane geometry, conics, circles, non-euclidean geometry (1922); Vol. iii. Solid geometry, quadrics, cubic curves in space, cubic surfaces (1923); Vol. iv. Higher geometry, being illustrations of the utility of the consideration of higher space, especially of four and five dimensions (1925).

GRASSMANN, H. Die lineale Ausdehnungslehre. Leipzig, 1844. 2. ed. 1878. Werke, Band i., Teil 1.

— Die Ausdehnungslehre. Berlin, 1862. Werke, Band i., Teil 2.

JESSOP, C. M. A treatise on the line complex. Cambridge: Univ. Press, 1903.

KLEIN, F. Über Liniengeometrie und metrische Geometrie. Math. Ann., 5 (1872), 257-277. Ges. Math. Abh., i, p. 106.

MÜLLER, E. Die verschiedenen Koordinatensysteme. Encykl. math. Wiss., III, 1_1 (III, AB 7). 1910.

PLÜCKER, J. Neue Geometrie des Raumes. Leipzig, 1868, 1869.

— Gesammelte wiss. Abh., I. Bd., Math. Abh. Leipzig, 1895.

CHAPTER VII

POLYTOPES

1. A polytope, the analogue of a polygon in two dimensions and a polyhedron in three dimensions, is a figure bounded by hyperplanes and specified more particularly in ways which we shall describe. Adjacent hyperplanes meet in boundaries of $n - 2$ dimensions, and in the usual types of polytopes two and only two hyperplanes meet in each boundary of $n - 2$ dimensions. Three or more adjacent $(n - 1)$-flats meet in boundaries of $n - 3$ dimensions; p or more meet in boundaries of $n - p$ dimensions, n or more meet in points, the *vertices*. For shortness we may call a boundary of r dimensions an r-boundary.

2. The Simplex. The simplest polytope in S_n is the simplex $S(n + 1)$, which is bounded by $n + 1$ hyperplanes. It is the analogue of a triangle and a tetrahedron. Every hyperplane cuts every other hyperplane in an $(n - 2)$-flat, and there are therefore ${}_{n+1}C_2 = \frac{1}{2}n(n + 1)$ boundaries of $n - 2$ dimensions, each an $S(n)$. The boundaries of $n - 3$ dimensions are formed by the intersections of sets of three hyperplanes, and so on. The vertices, $n + 1$ in number, are the intersections of all but one of the hyperplanes. Let N_r be the number of r-boundaries, and N_{pq} the number of p-boundaries which lie in (if $p < q$) or pass through ($p > q$) each q-boundary. Then

$$N_r = {}_{n+1}C_{r+1} = \frac{(n + 1)!}{(r + 1)!\,(n - r)!},$$

and

$$N_{pq} = {}_{q+1}C_{p+1} \qquad (p < q),$$

$$N_{pq} = {}_{n-q}C_{p-q} \qquad (p > q).$$

N_r is the co-efficient of x^{r+1} or x^{n-r} in the expansion of $(1 + x)^{n+1}$. Putting $x = -1$ we have

$$1 - N_0 + N_1 - \dots + (-)^n N_{n-1} + (-1)^{n+1} = 0.$$

This formula, which will be established later for a general class of polytopes, is the n-dimensional extension of *Euler's Polyhedral Formula* in three dimensions, viz. $N_0 - N_1 + N_2 = 2$. For two dimensions it reduces to $N_0 - N_1 = 0$ (the number of edges of a polygon is equal to the number of vertices), for four dimensions $N_0 - N_1 + N_2 - N_3 = 0$. It will be noticed that when n is even the formula is homogeneous in the N_r, but when n is odd there is a term independent of the N's; we shall deduce an important consequence of this later.

3. **Configurations.** The simplex is a particular example of a *configuration*. A configuration is a system of points, lines, planes, etc., such that through every point there pass the same number of lines, the same number of planes, etc.; on every line there are the same number of points, through every line the same number of planes, etc. A configuration can be represented by a square array or matrix of symbols N_{pq} denoting the number of p-flats which are incident with each q-flat, i.e. when $p > q$, N_{pq} denotes the number of p-flats passing through each q-flat, and when $p < q$, the number of p-flats lying in each q-flat. N_{pp}, which may be written also N_p, denotes the total number of p-flats in the configuration. The dimensions of the containing space being n, we have

$$N_p = N_{pp} = N_{pn} \quad . \quad . \quad . \quad (3\cdot1)$$

4. These numbers are not independent. If we reckon up the number of p-flats by counting N_{pq} at each of the N_q q-flats $(p > q)$ we obtain $N_q \, . \, N_{pq}$. But each p-flat is counted once for every q-flat which lies in it. Hence

$$N_q \, . \, N_{pq} = N_p \, . \, N_{qp} \quad . \quad . \quad . \quad (4\cdot1)$$

This relation is symmetrical in p, q and therefore holds whether $p >$ or $< q$.

We may also use the 3-index symbol ${}_rN_{pq}$ $(r > p > q)$, which denotes the number of p-flats passing through a given q-flat and lying in a given r-flat. If $r = n$ we have

$${}_nN_{pq} = N_{pq}. \quad . \quad . \quad . \quad (4\cdot2)$$

If $r > q > p$

$${}_rN_{pq} = N_{pq}. \quad . \quad . \quad . \quad (4\cdot3)$$

Also $${}_rN_{pp} = {}_rN_{pr} = N_{pr} \cdot \; (r > p). \quad . \quad . \quad (4\cdot4)$$

Writing the equation (4·1) in the form

$$ {}_nN_{qn} \,.\, {}_nN_{pq} = {}_nN_{pn} \,.\, {}_nN_{qp} $$

and changing n into r, we have (for $r > p > q$)

$$ N_{qr} \,.\, {}_rN_{pq} = N_{pr} \,.\, N_{qp} \quad . \quad . \quad . \quad (4·5) $$

Hence the 3-index symbols are all expressible in terms of the 2-index symbols.

5. A polytope is not in general a configuration. We may still use the symbols N_p with the same meaning, but N_{pq} will in general have different values for the different q-dimensional boundaries. The equations (4·1) can still be applied in a modified form. If there are $N_p^{(1)}$ boundaries of p dimensions which contain $N_{qp}^{(1)}$ q-boundaries, $N_p^{(2)}$ containing $N_{qp}^{(2)}$, etc., and $N_q^{(1)}$ boundaries of q dimensions through which pass $N_{pq}^{(1)}$ p-boundaries, etc., then we have

$$ N_q^{(1)}N_{pq}^{(1)} + N_q^{(2)}N_{pq}^{(2)} + \ldots = N_p^{(1)}N_{qp}^{(1)} + N_p^{(2)}N_{qp}^{(2)} + \ldots , $$

and also

$$ N_p^{(1)} + N_p^{(2)} + \ldots = N_p, $$
$$ N_q^{(1)} + N_q^{(2)} + \ldots = N_q. $$

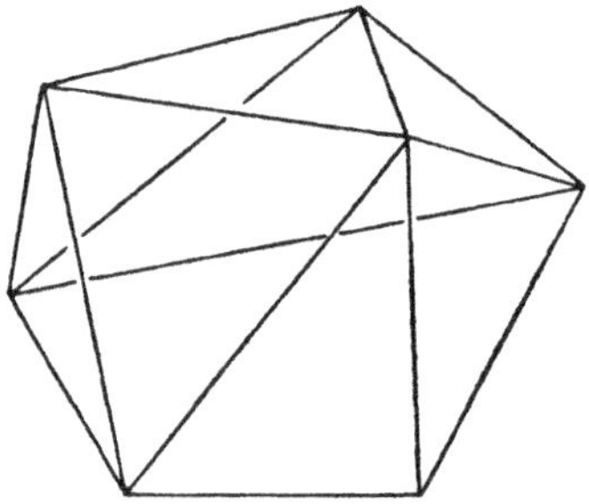

FIG. 12.

As an example consider the polyhedron (Fig. 12) in which

$$ N_2 = 9, \; N_1 = 14, \; N_0 = 7. $$

$N_{01} = N_{21} = 2$ for each of the 14 edges, N_{02} has the value 3 for 8 faces, and 4 for one. We can represent N_{02} by the symbol $3^8 4^1$. Then $\Sigma N_2 N_{02} = (8 \times 3) + 4 = 28$. Also N_{20} is of type $3^1 4^5 5^1$ and $\Sigma N_0 N_{20} = 3 + (5 \times 4) + 5 = 28$. N_{10} is of type $3^1 4^5 5^1$, the same as N_{20}, and $\Sigma N_0 N_{10} = 28 = N_1 N_{01}$.

6. Simple Polytopes. A polytope may be simple or complex. In a simple polytope two and only two $(n - 1)$-boundaries meet at each $(n - 2)$-boundary, and the same is true for the boundaries of any dimensions, viz. in any p-boundary two and only two $(p - 1)$-boundaries meet at each $(p - 2)$-boundary, i.e.

$$ {}_pN_{p-1,\,p-2} = 2 \quad (p = 2, 3, \ldots, n). $$

Also $N_{01} = 2$; this may be included in the general statement if we understand ${}_qN_{p,\,-1}$ to mean the number of p-boundaries, without restriction, which lie in a q-boundary, i.e. $N_{pq}(p < q)$, so that, for $p = 1$, ${}_pN_{p-1,\,p-2} = {}_1N_{0,\,-1} = N_{01}$.

An example of a complex polygon is the complete quadrilateral, which has four edges and six vertices. We shall confine our attention generally to simple polytopes.

Ex. Show that for any simple polyhedron

$$\Sigma N_0 \,.\, N_{10} = \Sigma N_0 \,.\, N_{20} = \Sigma N_2 \,.\, N_{02} = \Sigma N_2 \,.\, N_{12} = 2N_1.$$

7. Convex Polytopes. A polytope is *convex* when it lies entirely to one side of each of its $(n - 1)$-boundaries. If a polytope is convex so also are each of its boundaries of any dimensions. For suppose a polytope to have a p-boundary which is not convex, but is such that the $(p - 1)$-flat S_{p-1} containing one of its $(p - 1)$-boundaries divides the p-boundary, then an $(n - 1)$-flat containing S_{p-1} but not containing the p-flat divides the p-boundary in the same way; it therefore divides the polytope, and the polytope is not convex. Conversely, however, a polytope whose boundaries are all convex may not itself be convex.

8. Face and Vertex Constituents. We shall confine our attention generally to simple convex polytopes, and we shall denote, with Schoute, a simple convex polytope of n dimensions by $(\mathrm{Po})_n$.

Each $(n - 1)$-boundary of a $(\mathrm{Po})_n$ is a $(\mathrm{Po})_{n-1}$. We may call these the *face-constituents*. The configurational numbers ${}_rF_{pq}$ of a particular face-constituent are

$${}_rF_{pq} = {}_rN_{pq},$$

and in particular

$$F_p = N_{p,\,n-1},$$

where of course only one $(n - 1)$-boundary and only those r-boundaries which are contained in it are involved.

Consider any vertex O, and construct a small hypersphere with centre O. Each edge through O cuts the hypersphere in a point, the plane of each two-dimensional boundary cuts the hypersphere in a line (arc of great circle), and so on, and we have on the hypersphere a hyperspherical polytope of $n - 1$

dimensions. We shall call this a *vertex-constituent*. The configurational numbers ${}_rV_{pq}$ of a particular vertex-constituent are

$$ {}_rV_{pq} = {}_{r+1}N_{p+1,\ q+1}, $$

and in particular

$$ V_p = N_{p+1,\ 0}. $$

9. Isomorphism. Polytopes may be investigated fundamentally from the point of view of morphology, without respect to measurement, lengths of edges, or magnitudes of angles. This is part of the subject-matter of *analysis situs*. It is more general even than projective geometry, for we are not concerned, for example, whether the edges are straight or curved, or the faces flat; we are only concerned with the arrangement of the parts, i.e. the configurational numbers. When two polytopes are such that the vertices, edges, etc., can be made to correspond each to each, the edge joining a pair of vertices of the one corresponding to the edge joining the corresponding vertices of the other, and similarly for all their boundaries, they are said to be *isomorphic*. They have then the same set of numbers ${}_rN_{pq}$. Thus all simplexes of n dimensions are isomorphic. Any two simple quadrilaterals are isomorphic. The cube is isomorphic with any hexahedron whose faces are all quadrilaterals.

10. Schlegel Diagrams. A very useful method of representing a convex polyhedron is by a plane projection. If it is projected from any external point, since each ray cuts it twice, it will be represented by a polygonal area divided twice over into polygons. It is always possible by suitable choice of the centre of projection to make the projection of one face completely contain the projections of all the other faces. This is called a *Schlegel diagram* of the polyhedron. The Schlegel diagram completely represents the morphology of the polyhedron. It is sometimes convenient to project the polyhedron from a vertex; this vertex is projected to infinity and does not appear in the diagram, the edges through it are represented by lines drawn outwards. Thus a cube is represented by either of the figures in Fig. 13.

In four dimensions the Schlegel diagram or model of a polytope is a three-dimensional projection or representation in which the projection of one of its boundary polyhedra contains all the others. Thus the Schlegel diagram of a simplex in S_4 is a tetrahedron divided into four tetrahedra.

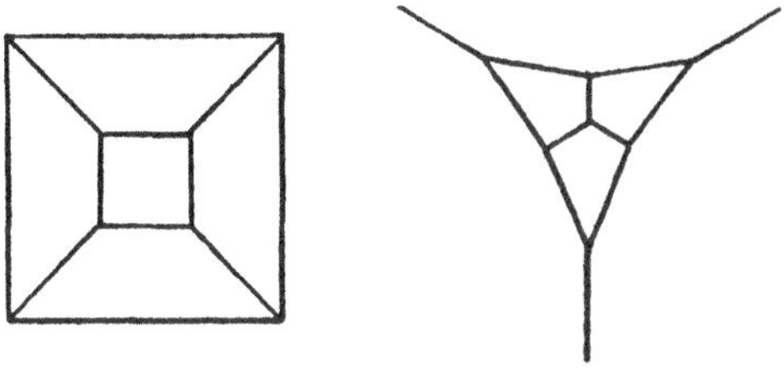

FIG. 13.—Schlegel diagrams of a cube.

11. Allomorphism. It is not sufficient for isomorphism that the two polytopes should have the same *total configurational numbers* N_p. Thus the polyhedron whose Schlegel diagram is represented in Fig. 14 has 6 faces, 12 edges, and 8 vertices, the same as a cube, and it has also three edges at each vertex, but its faces consist of 2 pentagons, 2 quadrilaterals, and 2 triangles. Two polytopes which have the same numbers N_p, but are not isomorphic, are said to be *allomorphic.*

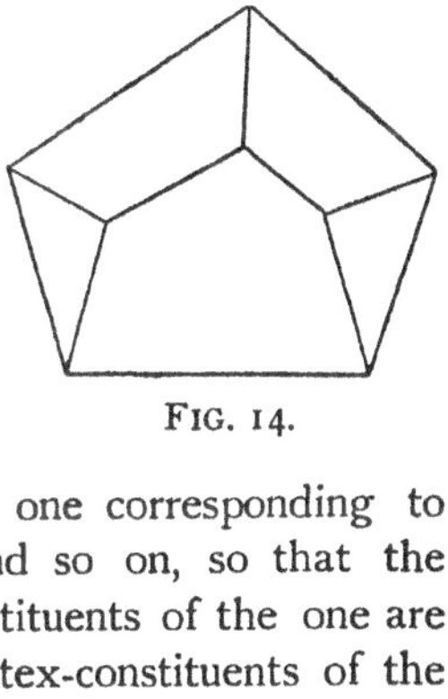

FIG. 14.

12. Reciprocal Polytopes. Again when two polytopes can be made to correspond reciprocally, the vertices of one corresponding to the $(n - 1)$-boundaries of the other, and so on, so that the configurational numbers of the face-constituents of the one are equal to those of the corresponding vertex-constituents of the other, they are said to be *reciprocal* or *polar-isomorphic.* The term reciprocal, in the strict sense, is applied when the two figures are related by a polarity with regard to a certain quadric or hypersphere.

The conditions that two polytopes should be reciprocal are

$$N'_{p-1,\, q-1} = N_{n-p,\, n-q}.$$

If $N_{p-1,\, q-1} = N_{n-p,\, n-q}$ the polytope is *self-reciprocal.*

If the conditions only hold for the total configurational numbers, $N'_{p-1} = N_{n-p}$, the polytopes are said to be *polar-*

allomorphic. A polytope may be *autopolar-allomorphic*, without being self-reciprocal. For example in the case of the polyhedron represented in Fig. 15.

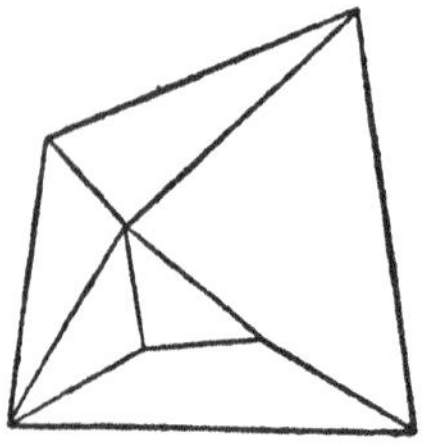

FIG. 15.

$$N_0 = 7,\ N_1 = 12,\ N_2 = 7;$$
but $N_{02} = 3^4 4^3$, while $N_{20} = 3^5 4^1 5^1$.

13. Triangular and Trihedral Polyhedra. The simplest general type of polyhedron from the point of view of its faces is one whose faces are all triangles. This may be called a *triangular polyhedron.* The simplest from the point of view of its vertices is one which has just three edges at each vertex, i.e. one whose vertex-constituents are all triangles. This may be called a *trihedral polyhedron.* The reciprocal of a triangular polyhedron is a trihedral polyhedron. Any polyhedron can be converted into a triangular polyhedron by dividing each face into triangles, or by erecting pyramids on each of those faces which have more than three sides; and into a trihedral polyhedron by truncating every vertex at which there are more than three edges. These two processes, *truncating* and *pointing*, are reciprocal.

14. Simplex-Polytopes and Simplex-Polycoryphas. To distinguish between a polytope considered with respect to its faces and its vertices, we may call it in the latter case a *polycorypha.**

The simplest general type of polytope is one whose $(n - 1)$-boundaries are all simplexes. This may be called a *simplex-polytope.* Every boundary of p dimensions of a simplex-polytope is a simplex $S(p + 1)$. The simplest general type of polycorypha is one whose vertex-constituents are all simplexes, i.e. having n edges at each vertex. This may be called a *simplex-polycorypha.* In a simplex-polycorypha there are $n - p$ boundaries of $n - 1$ dimensions through every p-boundary. The process of truncating a polytope at all those vertices which contain more than n edges replaces each vertex by a simplex $S(n)$ and produces a simplex-polycorypha. The pro-

* κορυφή is Euclid's name for the vertex of a polyhedron. Cayley uses the term *polyacron.*

cess of pointing a polytope consists in joining each of the vertices of an $(n - 1)$-boundary to an external point near that boundary, i.e. erecting an n-dimensional pyramid on the face. If this is done for every face which has more than n boundaries the figure becomes a simplex-polytope.

In truncating it has to be noticed, of course, that the cutting hyperplane must cut only the edges which proceed from the given vertex. Deeper sections will lead to more complicated figures.

15. Sections and Frusta of a Simplex. If the boundaries of the simplex are considered as indefinitely extended, any hyperplane will cut the figure in a complete $(n + 1)$-S_{n-2} in S_{n-1}, analogous to a complete quadrilateral or 4-S_1. If, however, we consider only the interior parts of the boundaries the section is a polytope of one dimension less than the simplex. Also the hyperplane divides the simplex into two separate polytopes, *frusta* of the simplex. The forms of the section and the frusta depend upon the position of the hyperplane with regard to the vertices of the simplex, and there is a close connection between the form of the section and those of the frusta. If the hyperplane separates the $n + 1$ vertices of the simplex into two groups p and $n + 1 - p$ we shall denote the type of the polytope of section by $(p, n + 1 - p)$ so that, of course, $(p, n + 1 - p) = (n + 1 - p, p)$; and the types of the polytopes into which it is divided will be denoted by $(p \mid n + 1 - p)$ and $(n + 1 - p \mid p)$, the first of the two numbers in the symbol denoting the number of vertices of the simplex which the frustum contains.

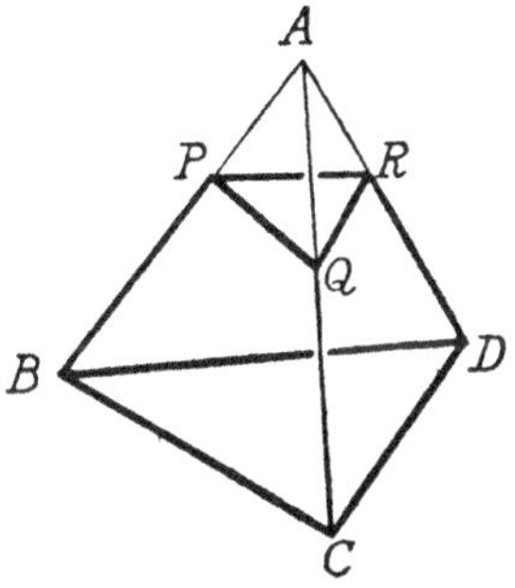

FIG. 16.

16. Let us consider first a tetrahedron ABCD (Fig. 16), and first let the plane of section divide the vertices into groups 1, 3. The section is a triangle PQR, i.e. (1, 3) denotes a triangle. The frusta are of type $(1 \mid 3)$ a tetrahedron, and $(3 \mid 1)$ a pentahedron bounded by 2 triangles and 3 quadrilaterals. (This pentahedron, which is isomorphic with a triangular prism, is, as it happens, the only type of convex pentahedron having 3 faces at each vertex. See § 20.)

Second, let the section divide the vertices into groups 2, 2 (Fig. 17). The section, of type (2, 2), is a quadrilateral, and the two frusta are of the same type (2 | 2), again a pentahedron, i.e. (2 | 2) = (3 | 1).

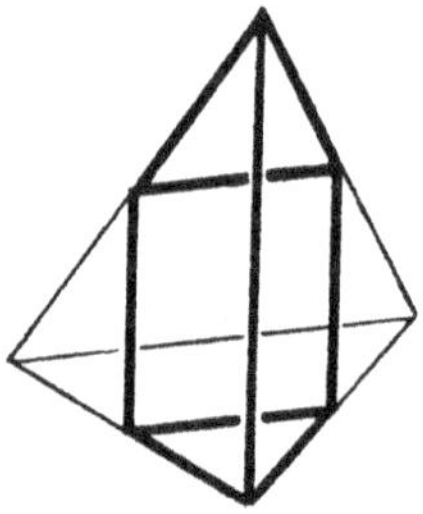

FIG. 17.

17. Consider next a simplex ABCDE in S_4, and let the hyperplane of section divide the 5 vertices into groups 1, 4, say A and BCDE. The vertices of the section are the points of intersection with the lines AB, AC, AD, AE; the edges are the lines of intersection with the planes ABC, ABD, ABE, ACD, ACE, ADE; and the faces are the planes of intersection with the hyperplanes ABCD, ABCE, ABDE, ACDE. Thus the section (1, 4) is a tetrahedron. The frustum (1 | 4) is a simplex, the frustum (4 | 1) is a polytope bounded by the section (a tetrahedron), the tetrahedron BCDE, and four polyhedra (3 | 1), the frusta of the tetrahedra ABCD, etc.

Next let the hyperplane divide the vertices into groups 2, 3, say AB and CDE. The vertices of the section are the points of intersection with the lines AC, AD, AE, BC, BD, BE; the edges are the lines of intersection with the planes ABC, ABD, ABE, ACD, ACE, ADE, BCD, BCE, BDE; and the faces are the planes of intersection with the hyperplanes ABCD, ABCE, ABDE, ACDE, BCDE. The first three of these faces are (2, 2) sections of tetrahedra, i.e. quadrilaterals, and the last two are (1, 3) sections, i.e. triangles. Hence the section is a pentahedron (3 | 1) or (2 | 2). The frustum (2 | 3) is a polytope bounded by the section (a pentahedron), and the frusta of the tetrahedra ABCD, etc.; three of these, ABCD, ABCE, ABDE, are of type (2 | 2), i.e. pentahedra, and two, ACDE, BCDE, of type (1 | 3), i.e. tetrahedra. (2 | 3) is therefore a polytope bounded by 2 tetrahedra and 4 pentahedra. Hence (2 | 3) = (4 | 1). The frustum (3 | 2) is bounded by the section (a pentahedron), three frusta of type (2 | 2) (pentahedra), and two of type (3 | 1) (also pentahedra). (3 | 2) is therefore a polytope bounded by 6 pentahedra. (See Fig. 22, p. 108.)

18. The numbers of boundaries of a section of a simplex in the general case are easily found. Consider a section of type (p, q), i.e. the section of a simplex in space of $p + q - 1$ dimensions by a hyperplane which separates the vertices into two groups p and q. We obtain an r-boundary by intersection with an $(r + 1)$-flat; this is determined by $r + 2$ points, and at least one of these points must be taken from each of the two groups. Hence

$$N_r = {}_{p+q}C_{r+2} - {}_pC_{r+2} - {}_qC_{r+2}.$$

A section by a hyperplane passing between the vertex and the base of the simplex, i.e. a section of type $(1, n)$ is a simplex. Any other section cuts all the $n + 1$ hyperplanes of the simplex and is a polytope with $n + 1$ cells.

The nature of the various cells, and other boundaries, is also at once found. The r-boundaries consist of ${}_pC_1 \cdot {}_qC_{r+1}$ sections of type $(1, r + 1)$, ${}_pC_2 \cdot {}_qC_r$ sections of type $(2, r)$, and in general ${}_pC_s \cdot {}_qC_{r+2-s}$ sections of type $(s, r + 2 - s)$. The cells, or $(p + q - s)$-boundaries, consist of ${}_pC_s \cdot {}_qC_{p+q-1-s}$ sections of type $(s, p + q - 1 - s)$ for $s = 1, 2, \ldots, p+q-1$.

19. A Frustum of a Simplex of $p + q$ Dimensions of Type $(p \mid q)$ is Isomorphic with a Section of a Simplex of

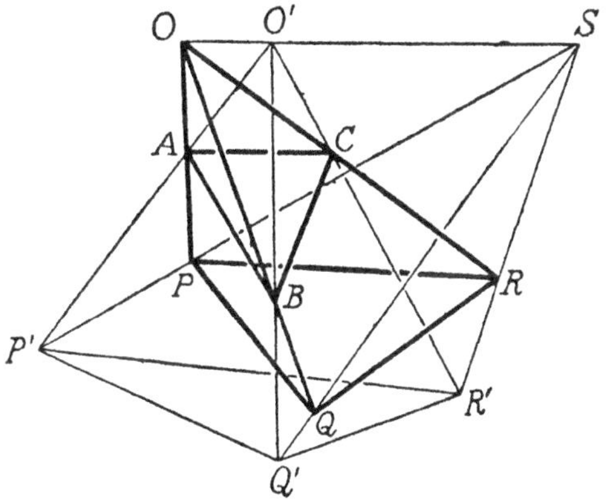

FIG. 18.

$p + q + 1$ Dimensions of Type $(p, q + 1)$. Consider first (Fig. 18) a frustum ABCPQR of a tetrahedron OPQR of type

$(3 \mid 1)$. Take any point S outside the space S_3 of the tetrahedron, and in SO take any point O′. SP, O′A, cut in P′; SQ, O′B in Q′; SR, O′C in R′. Then the simplex SO′P′Q′R′ in S_4 is cut by S_3 in the section ABCPQR, which is of type $(3, 2)$ since the space S_3 separates the vertices S, O′ from the vertices P′, Q′, R′.

Now consider a frustum $(p \mid q)$ of a simplex in S_{p+q-1}. Let the vertices of the simplex be $A_1, \ldots, A_p, B_1, \ldots, B_q$, and those of the section P_{ij} $(i = 1, \ldots, p;\ j = 1, \ldots, q)$, the vertex P_{ij} being the intersection of the line A_iB_j with the cutting hyperplane S. Take any point O not in the hyperspace S_{p+q-1} of the simplex. Join O to A_i, and between O and A_i take any point A'_i. Let A'_1P_{1j} cut OB_j in B'_j. We thus obtain a simplex $OA'_1 \ldots A'_pB'_1 \ldots B'_q$ in S_{p+q}. The points A_i are all outside this simplex, while B_i lie on its edges, and the simplex is cut by S_{p+q-1} in the section $P_{11}\, P_{12} \ldots P_{pq}B_1 \ldots B_q$, which is the frustum $(q \mid p)$ of the simplex $A_1 \ldots A_pB_1 \ldots B_q$. The points A'_i and B'_j are separated by P_{ij} and therefore lie on opposite sides of the section, and since O is separated from B'_j by B_j, O lies on the opposite side of the section from B'_j, and therefore on the same side as A'_i. The section is therefore of type $(q, p+1)$. Hence $(q \mid p) = (q, p+1) = (p+1, q) = (p+1 \mid q-1)$.

20. Enumeration of Polytopes. An immediate result of this is that in S_n there are $[\frac{1}{2}n]$ different types of simplex-polycoryphas with $n+2$ cells, i.e. $\frac{1}{2}n$ or $\frac{1}{2}(n-1)$ according as n is even or odd, the polytope being obtained as a frustum of a simplex in S_n or a section of a simplex in S_{n+1}, excluding the section of type $(1, n+1)$ which is a simplex.

Reciprocally there are $[\frac{1}{2}n]$ different types of simplex-polytopes with $n+2$ vertices.

Thus in S_3 there is only one type of trihedral polyhedron with 5 faces, viz. the pentahedron isomorphic with the frustum of a tetrahedron; it is bounded by 2 triangles and 3 quadrilaterals (Fig. 19). Reciprocally there is only one type of triangular polyhedron with 5 vertices (Fig. 20).

There is one other pentahedron, viz. the quadrilateral pyramid, and this is self-reciprocal.

Ex. Prove that every pyramid in S_3 is self-reciprocal.

In S_4 there are two types of simplex-polycoryphas with 6 cells, viz. the sections of type (2, 4) and (3, 3) of a simplex in S_5.

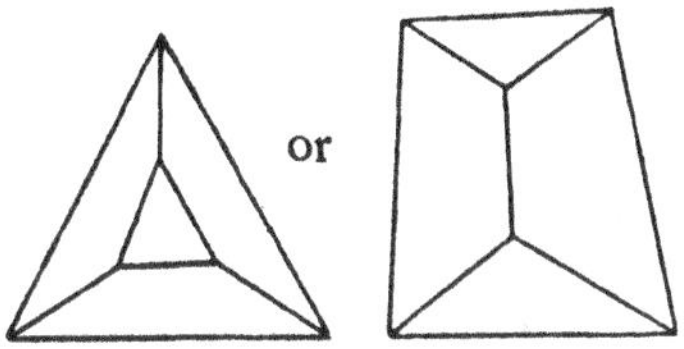

FIG. 19.—The trihedral pentahedron.

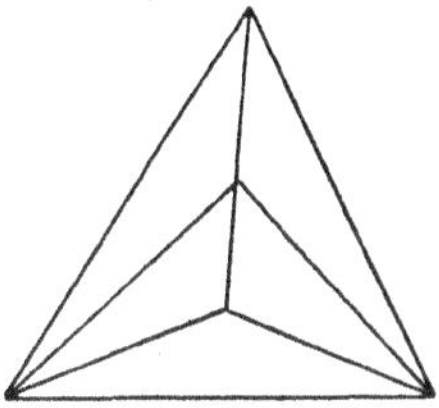

FIG. 20.—The triangular polyhedron with 5 vertices.

Denoting the vertices of the simplex in S_5 by A_1, A_2, B_1, B_2, B_3, B_4, the cells of the four-dimensional section are sections of the following simplexes:

2 A's and 3 B's: type (2, 3), a pentahedron,
1 A and 4 B's: type (1, 4), a tetrahedron.

The section of type (2, 4) is therefore bounded by 2 tetrahedra and 4 pentahedra.

For type (3, 3), denoting the vertices of the simplex in S_5 by A_1, A_2, A_3, B_1, B_2, B_3, the cells are sections of the following simplexes:

3 A's and 2 B's: type (3, 2), a pentahedron,
2 A's and 3 B's: type (2, 3), a pentahedron.

The section of type (3, 3) is therefore bounded by 6 pentahedra.

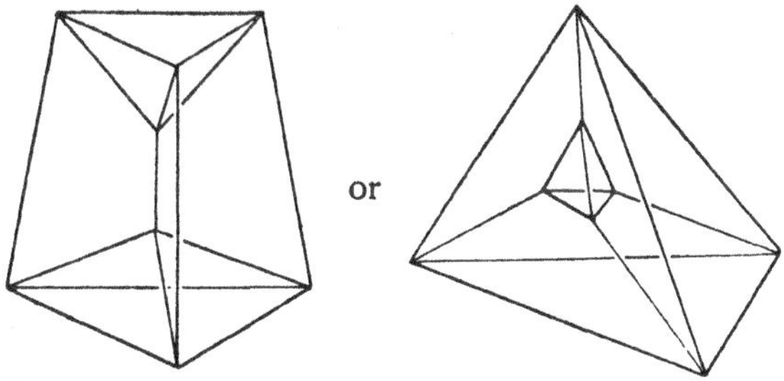

FIG. 21.—Simplex polycorypha bounded by 2 tetrahedra and 4 pentahedra.

The Schlegel diagrams of these two types of simplex-polycoryphas with 6 cells in S_4 are represented in Figs. 21 and 22.

The reciprocals are simplex-polytopes with 6 vertices, one with 8 cells, the other with 9 (Figs. 23 and 24).

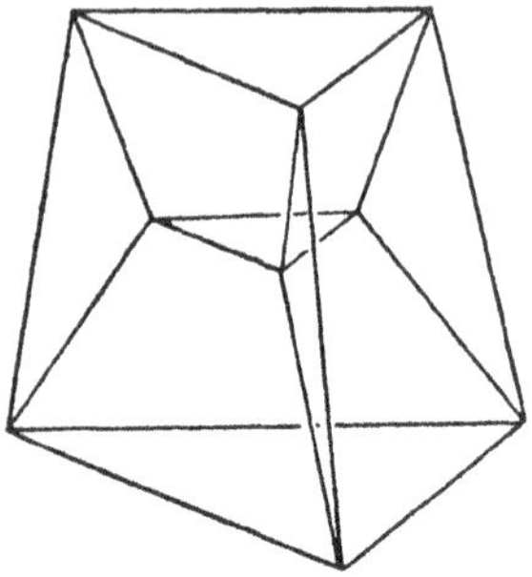

FIG. 22.—Simplex-polycorypha bounded by 6 pentahedra.

21. There are two other forms of polytopes in S_4 with 6 cells. To determine these we notice first that no cell can have more than 5 faces, since there must be a cell on each face, and hence the cells are tetrahedra, trihedral pentahedra, or quadrilateral pyramids. Also there can only be 4 or 5 cells at a vertex. We have determined those with 4 cells at each vertex. If there is a vertex at which 5 cells meet, the section near this vertex is either the trihedral pentahedron or the quadrilateral pyramid.

In the former case there are 6 edges at the vertex, and of the 5 cells two have trihedral vertices and three have tetra-

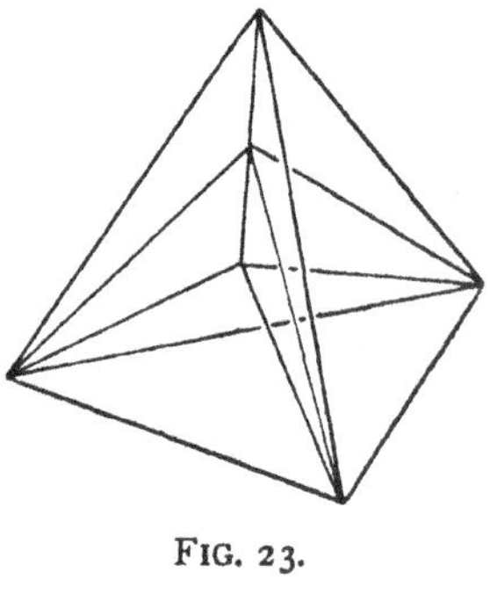

FIG. 23.

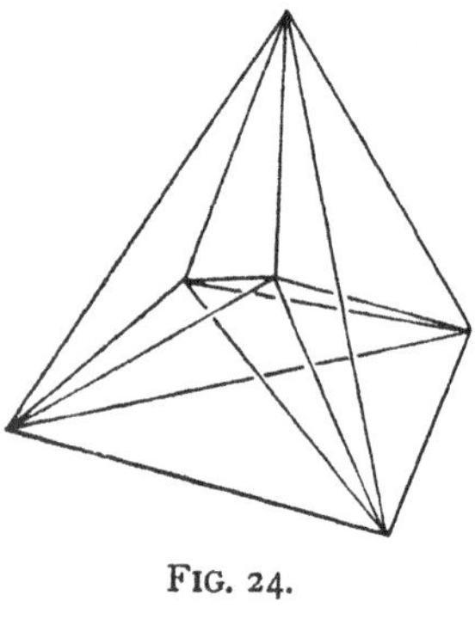

FIG. 24.

hedral, the latter being therefore quadrilateral pyramids. The polytope is then a pyramid formed by joining an external point to the vertices of a trihedral pentahedron (Fig. 25).

In the other case there are just 5 edges at the vertex, and of the 5 cells four have trihedral vertices and one tetrahedral, the latter being a quadrilateral pyramid. The polytope is then a pyramid with a quadrilateral pyramid as base (Fig. 26).

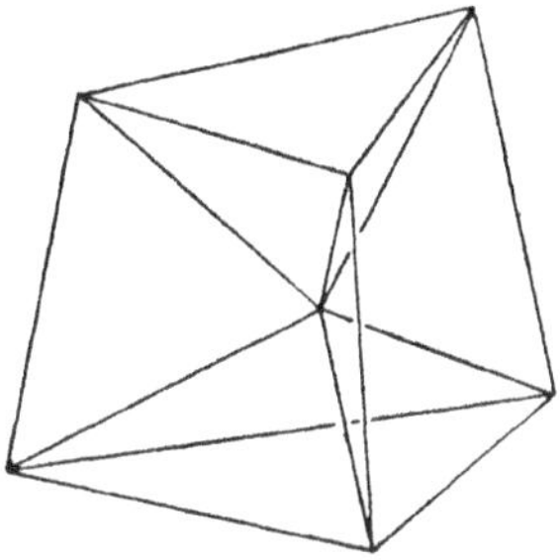

FIG. 25.

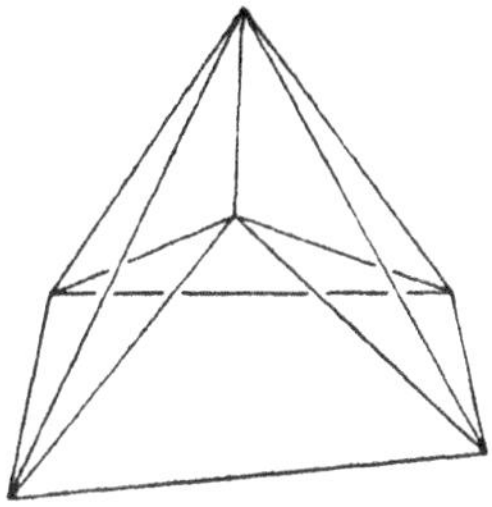

FIG. 26.

22. All the different types of trihedral polyhedra with a given number N of faces can be obtained by taking a tetrahedron along with N − 4 other planes, observing the condition that not more than three planes pass through one point.

A single additional plane gives a pentahedron, of which we have seen there is only one form.

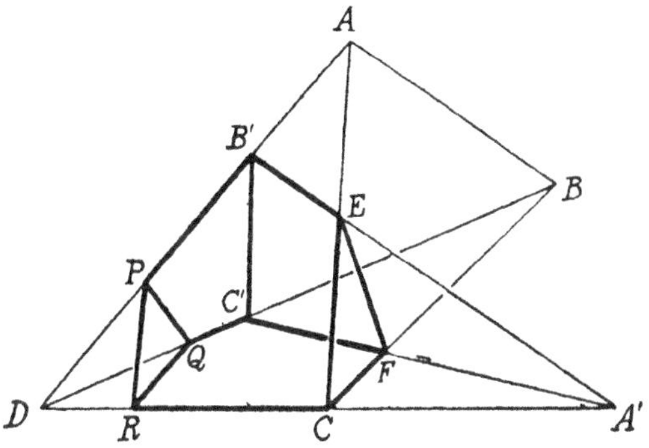

AB, C, D cut (2, 1, 1)
A′, B′C′, D cut (1, 2, 1) } $3^2 4^2 5^2$.

FIG. 27.

Two additional planes give rise to several cases, first according as the line of intersection of the two planes does or does not cut the tetrahedron, and second according to the disposition of the vertices of the tetrahedron in the spaces into which the planes divide the tetrahedron.

Thus if the line of intersection lies entirely outside the tetrahedron we have (retaining of course only the cases in which each plane separates the vertices of the tetrahedron) the cases in which the vertices have the dispositions 3, 0, 1 ; 2, 1, 1 ; 2, 0, 2 ; 1, 2, 1. The first case gives just a pentahedron. The third gives the type 4^6 (Figs. 28 and 29) (6 quadrilateral faces),

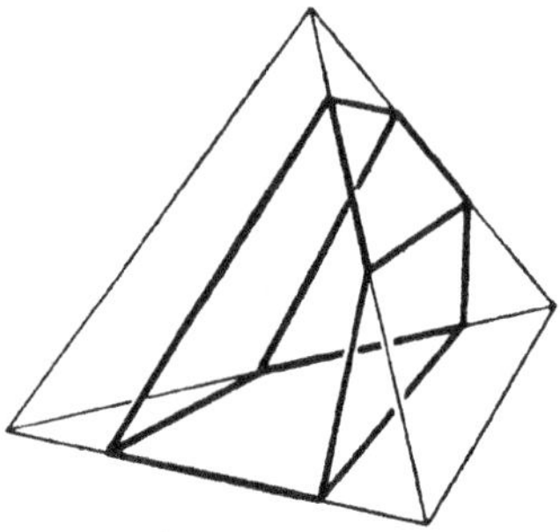

FIG. 28.—(2, 0, 2) 4^6.

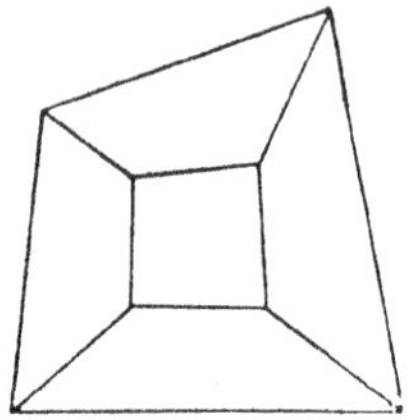

FIG. 29.—4^6.

isomorphic with a cube. The second and fourth lead to the same type $3^2 4^2 5^2$ (2 triangles, 2 quadrilaterals, and 2 pentagons), as is seen by considering the tetrahedron ABCD with the sections B′EFC′ and PQR, and the tetrahedron A′B′C′D with the sections ECF and PQR (Figs. 27 and 30). For both polyhedra $N_2 = 6$, $N_1 = 12$, $N_0 = 8$; i.e. they are allomorphic.

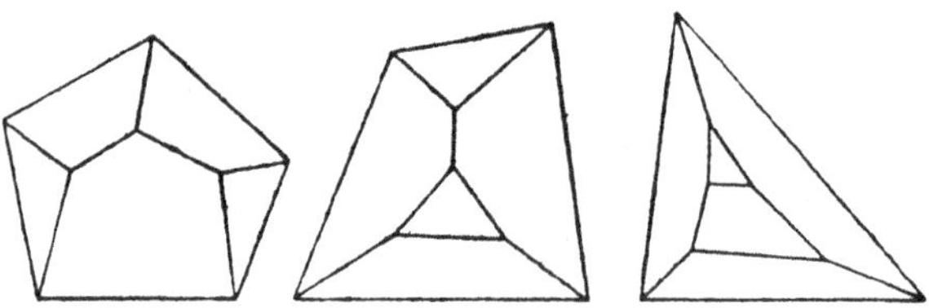

FIG. 30.—$3^2 4^2 5^2$.

When the line of intersection of the two cutting planes cuts the tetrahedron we get again various cases, but it is found that only the two foregoing types of polyhedra are produced.

The reciprocals of these form two types of triangular polyhedra with six vertices, one of which is isomorphic with the regular octahedron (Figs. 31 and 32).

In addition to the two trihedral hexahedra there are 5 other hexahedra (Fig. 33).

In a similar way the different types of simplex-polycoryphas

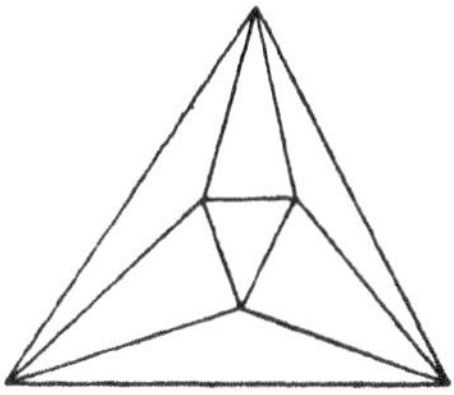

FIG. 31.

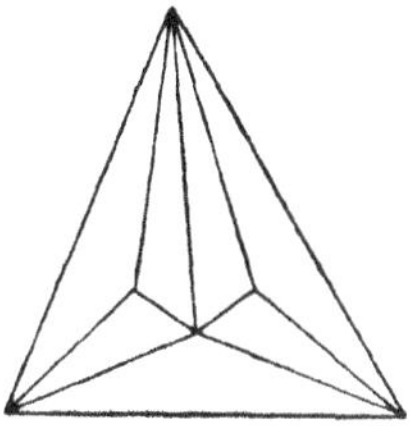

FIG. 32.

having $n + 3$ cells of $n - 1$ dimensions could be investigated by cutting a simplex with two hyperplanes. The problem of determining the different types of polytopes with a given

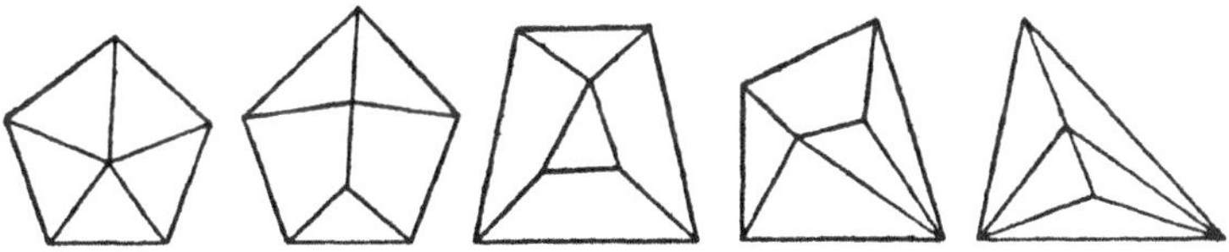

$3^5 5^1$. $3^3 4^2 5^1$. $3^2 4^4$. $3^4 4^2$. 3^6.

FIG. 33.

number N of boundaries becomes of great complexity as n and N increase, and we shall not pursue it further. The following short table gives the number of convex polyhedra in S_3 with N faces, from N = 4 to N = 8.

N.	4	5	6	7	8
Trihedral polyhedra . .	1	1	2	5	14
Total number of polyhedra	1	2	7	34	257

Polytopes of Special Form

23. Pyramids. A pyramid (of the first species) is constructed by joining all the points on the boundary of a convex polytope of $n - 1$ dimensions (the base) to a fixed point O (the vertex), not in the hyperplane of the base. If N'_p are the total configurational numbers of the base, those of the pyramid are

$$N_p = N'_p + N'_{p-1}, \quad (p = 1, 2, \ldots, n-2)$$
$$N_{n-1} = 1 + N'_{n-2}, \qquad N_0 = N'_0 + 1.$$

It is bounded by the base and N'_{n-2} pyramids of $n - 1$ dimensions.

If the base itself is a pyramid (of first species) with vertex A_2, the pyramid in S_n with vertex A_1 is said to be of the second species, and may be considered as a pyramid in two different ways. If α is the base of the $(n - 1)$-dimensional pyramid, the pyramid of the second species is constructed by forming first either the pyramid $A_1\alpha$ or the pyramid $A_2\alpha$, and then using this as base and the remaining point A_2 or A_1 as vertex. In addition to the boundaries of the base-polytope α, the pyramid of second species has as boundaries the vertices A_1, A_2; the edges A_1A_2 and those joining A_1 and A_2 to the vertices of α; the triangles formed by A_1, A_2 and a vertex of α, and A_1 or A_2 with an edge of α; and so on. If N'_p are the total configurational numbers of α, those of the pyramid of the second species are

$$N_p = N'_p + 2N'_{p-1} + N'_{p-2} \quad (p = 0, 1, \ldots, n-1)$$

with the understanding that

$$N'_{-1} = 1, \ N'_{-2} = 0, \ N'_{n-2} = 1, \ N'_{n-1} = 0.$$

In general a pyramid of species r is formed by joining the r vertices $A_1, \ldots, A_r$ of a simplex $S(r)$ to the boundary points of a polytope $(Po)_{n-r}$. If N'_p are the total configurational numbers of the base-polytope, those of the pyramid are

$$N_p = N'_p + rN'_{p-1} + {}_rC_2N'_{p-2} + \ldots + N'_{p-r},$$

it being understood that $N'_{-1} = 1 = N'_{n-r}$, while

$$N'_{-q} = 0 = N'_{n-r+q-1} \text{ if } q > 1.$$

In S_n a pyramid of species $n + 1$ is a simplex, and there is no pyramid of species n or $n - 1$. A triangle is a pyramid of species 3; a tetrahedron is a pyramid of species 4, any other pyramid in S_3 is of species 1.

Ex. Prove that a pyramid is self-reciprocal when and only when its base is self-reciprocal.

24. Prisms. A prism (of the first species) is generated by the parallel motion of a polytope $(\mathrm{Po})_{n-1}$; it is bounded by the polytope in its initial and final positions and by N'_{n-2} prisms of $n - 1$ dimensions. If N'_p are the total configurational numbers of the base-polytope, those of the prism are

$$\mathrm{N}_p = 2\mathrm{N}'_p + \mathrm{N}'_{p-1} \quad (p = 1, 2, \ldots, n-2)$$
$$\mathrm{N}_0 = 2\mathrm{N}'_0, \quad \mathrm{N}_{n-1} = 2 + \mathrm{N}'_{n-2}.$$

If the base-polytope is itself a prism (of the first species), the prism generated is one of the second species, and so on. A prism of species r is generated by a $(\mathrm{Po})_{n-r}$ moving in r independent directions. If N'_p are the total configurational numbers of the base-polytope, those of the prism are

$$\mathrm{N}_p = 2^r\mathrm{N}'_p + {}_r\mathrm{C}_1 2^{r-1}\mathrm{N}'_{p-1} + {}_r\mathrm{C}_2 2^{r-2}\mathrm{N}'_{p-2} + \ldots + \mathrm{N}'_{p-r},$$

where $\mathrm{N}'_{n-r} = 1$, while $\mathrm{N}'_{-q} = 0 = \mathrm{N}'_{n-r+q}$ if $q > 0$.

In S_n a prism of species n is a parallelotope, and there is no prism of species $n - 1$. A parallelogram is a prism of species 2, a parallelepiped is a prism of species 3. The total configurational numbers of a parallelotope are

$$\mathrm{N}_p = {}_n\mathrm{C}_p 2^{n-p}.$$

When the n directions are all mutually orthogonal we have the analogue of a rectangular solid, which we have called an *orthotope*. If, further, the displacements are all equal in length we have a regular polytope, the analogue of the cube.

25. The Simplotope. In three dimensions a triangular prism can be considered as generated in two ways. Take a tetrahedron OABC and let the face OAB move always parallel to itself so that O moves along OC, thus generating the triangular prism OABCA′B′; or alternatively let the edge OC move always parallel to itself so that O moves over the whole triangle OAB.

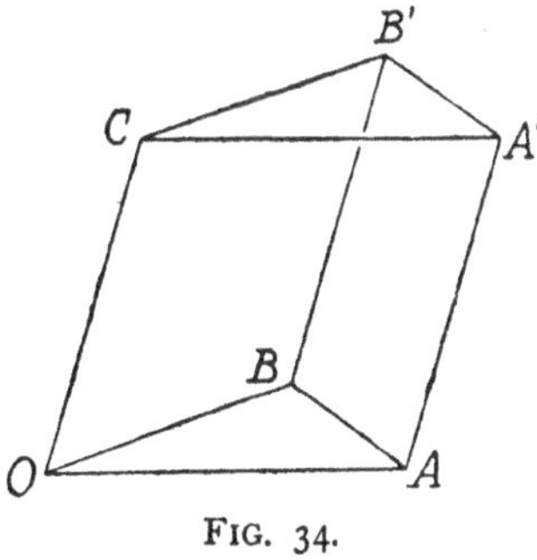

FIG. 34.

Generalising this, consider a simplex $\mathrm{OA}_1 \ldots \mathrm{A}_p\mathrm{B}_1 \ldots \mathrm{B}_q$

in S_{p+q}. Then a new polytope is generated by moving the simplex of p dimensions $P \equiv (OA_1 \ldots A_p)$ always parallel to itself so that O moves over the whole boundary of the simplex $Q = (OB_1 \ldots B_q)$, and the same polytope is generated when the rôles of the two simplexes are interchanged. This polytope is called a *simplotope* of type (p, q).

Number of Boundaries of a Simplotope (p, q) *in* S_{p+q}. When the point O of the generating simplex P is at a vertex of Q we obtain $p + 1$ vertices of the figure, hence $N_0 = (p + 1)(q + 1)$. Each vertex of P generates an edge when O moves along an edge of Q, and in each of its various positions when O is at a vertex of Q it furnishes ${}_{p+1}C_2$ edges, therefore $N_1 = {}_{p+1}C_1 \cdot {}_{q+1}C_2 + {}_{p+1}C_2 \cdot {}_{q+1}C_1$. Similarly each edge of P generates a parallelogram when O moves along an edge of Q, each vertex of P generates a triangle when O moves over a triangular boundary of Q, and in each position of P when O is at a vertex of Q it furnishes ${}_{p+1}C_3$ triangles. Hence

$$N_2 = {}_{p+1}C_2 \cdot {}_{q+1}C_2 + {}_{p+1}C_1 \cdot {}_{q+1}C_3 + {}_{p+1}C_3 \cdot {}_{q+1}C_1$$

of which ${}_{p+1}C_2 \cdot {}_{q+1}C_2$ are parallelograms and the rest triangles. In general

$$N_r = {}_{p+1}C_{r+1} \cdot {}_{q+1}C_1 + {}_{p+1}C_r \cdot {}_{q+1}C_2 + \ldots + {}_{p+1}C_1 \cdot {}_{q+1}C_{r+1},$$
$$N_{p+q-1} = {}_{p+1}C_{p+1} \cdot {}_{q+1}C_q + {}_{p+1}C_p \cdot {}_{q+1}C_{q+1}$$
$$= (q + 1) + (p + 1) = p + q + 2.$$

The boundaries of all dimensions are simplotopes. Of the r-boundaries ${}_{p+1}C_{s+1} \cdot {}_{q+1}C_{r-s+1} + {}_{p+1}C_{r-s+1} \cdot {}_{q+1}C_{s+1}$ are simplotopes of type $(s, r - s)$.

N_r is equal to the coefficient of x^{r+2} in the expansion of $(1 + x)^{p+1}(1 + x)^{q+1}$, i.e. $(1 + x)^{p+q+2}$, omitting the two terms ${}_{p+1}C_{r+2}$ and ${}_{q+1}C_{r+2}$, hence

$$N_r = {}_{p+q+2}C_{r+2} - {}_{p+1}C_{r+2} - {}_{q+1}C_{r+2}.$$

Comparing this with the result for a section of a simplex (p. 105) we find that the simplotope of type (p, q) is isomorphic with the section of type $(p + 1, q + 1)$ of a simplex in S_{p+q+1}.

If the vertices of a simplex in S_{p+q+1} are divided into two groups $p + 1$ and $q + 1$, these form two simplexes in flat spaces α and β of p and q dimensions respectively. These may be called *complementary boundaries* of the simplex. The spaces at infinity α_∞ and β_∞, of dimensions $p - 1$ and $q - 1$, determine a $(p + q - 1)$-flat at infinity. Hence there is a single infinity of $(p + q)$-flats completely parallel to both α and β. A simplotope is a section of a simplex by a hyperplane parallel to two complementary boundaries.

26. Pyramidoids. A pyramid of species r can be generalised further by replacing the simplex $S(r)$ by a polytope $(Po)_{r-1}$. Let $(Po)_p$ and $(Po)_q$ be two polytopes in S_{p+q+1}, so that their hyperplanes do not intersect. Then a polytope is formed by joining all the boundaries of the one polytope to all the boundaries of the other. We may call this a *Pyramidoid* of type $\{(Po)_p, (Po)_q\}$. Its other boundaries are also pyramidoids which may simplify to pyramids or simplexes. If $(Po)_p$ and $(Po)_q$ are both simplex-polytopes all the other boundaries of the pyramidoid are simplexes. Let N'_r and N''_r be the total configurational numbers of the two polytopes, and N_r those of the pyramidoid, then

$$N_r = N'_r + N'_{r-1}N''_0 + N'_{r-2}N''_1 + \ldots + N'_0N''_{r-1} + N''_r,$$

with the understanding that

$$N'_{-1} = 1 = N'_p, \quad N''_{-1} = 1 = N''_q,$$

while $\quad N'_{-s} = 0 = N'_{p+s-1}, \quad N''_{-s} = 0 = N''_{q+s-1} \quad (s > 1).$

In particular
$$N_{p+q} = N'_{p-1} + N''_{q-1},$$
$$N_0 = N'_0 + N''_0.$$

Ex. Show that the section of a pyramidoid with base polytopes $(Po)_{n-r}$ and $(Po)_{r-1}$ is isomorphic with the polytope generated by moving the polytope $(Po)_{r-1}$ always parallel to itself so that one of its vertices moves over the whole boundary of the polytope $(Po)_{n-r}$.

27. Prismoids. This process may be extended still further by removing the restriction that the hyperplanes α and β of the two base-polytopes $(Po)_p$ and $(Po)_q$ are skew. We shall only assume that they have no finite point in common, i.e. they are either skew or parallel, and may lie in a space S_n $(n \gtrless p + q + 1)$. If the condition of parallelism also is removed isomorphic polytopes of more general form result.

This class of polytopes, called *Prismoids*, may be represented generally by those constructed on a basis of two polytopes P′, P″, each of $n - 1$ dimensions, lying in parallel hyperplanes. One or both of these may degenerate to a polytope of lower dimensions. Assuming that P′ and P″ are both convex, a definite procedure for obtaining a convex prismoid is then as follows. Choose any $(n - 2)$-boundary of P′, then through this there are just two $(n - 1)$-flats which meet P″ in just a vertex; * one of these separates points of P′ and P″, the other lies entirely to one side of both. Choose the latter and form the pyramid with this boundary of P′ as base and the vertex of P″ as vertex. In this way we obtain N'_{n-2} pyramids as boundaries of the prismoid, and similarly we get N''_{n-2} pyramids with $(n - 2)$-boundaries of P″ as base and a vertex of P′ as vertex. Then taking an $(n - 3)$-boundary of P′, if the pyramids on the two $(n - 2)$-boundaries which pass through this $(n - 3)$-boundary do not have a common boundary of $n - 2$ dimensions, we form one or more pyramids of the second species with this $(n - 3)$-boundary as base and two vertices of P″ as vertices; and in general a pyramidoid as boundary is obtained with its bases a $(p - 1)$-boundary of P′ and an $(n - p)$-boundary of P″. If P′ and P″ are both simplex-polytopes, all the other boundaries of the prismoid are simplexes.

Ex. 1. Show that a prismoid in S_4 whose bases are two tetrahedra is a polytope bounded by 16 tetrahedra (the tetrahedral 16-cell), and that a section parallel to the base is a polyhedron bounded by 6 quadrilaterals and 8 triangles.

Ex. 2. The prismoid whose bases are a cube and an octahedron has for its remaining boundaries 6 quadrilateral pyramids and 20 tetrahedra. A section parallel to the base is a polyhedron bounded by 18 quadrilaterals and 8 triangles.

Ex. 3. If one base is a cube and the other a tetrahedron with each edge parallel to a corresponding face-diagonal of the

* It may happen that an $(n - 1)$-flat which is made to pass through a vertex of P″ contains as well an edge or higher boundary through this vertex; in this case some of the boundaries which would in general be separate lie in one flat and form a more complex single boundary.

cube, the remaining boundaries of the prismoid are 4 tetrahedra, and 6 prismoids with base a square and top a single edge.

Ex. 4. A prismoid whose bases are a cube and a tetrahedron in general position has for its remaining boundaries 6 quadrilateral pyramids and 16 tetrahedra. A section parallel to the base is a polyhedron bounded by 18 quadrilaterals and 4 triangles; 42 edges and 22 vertices.

REFERENCES

BRÜCKNER, M. Vielecke und Vielflache. Theorie und Geschichte. Leipzig, 1900.

— Die Elemente der vierdimensionalen Geometrie mit besonderer Berücksichtigung der Polytope. Zwickau, Jahresber. Ver. Natk., 1893, pp. 61.

EBERHARD, V. Zur Morphologie der Polyeder. Leipzig, 1891.

SCHLEGEL, V. Theorie der homogenen zusammengesetzten Raumgebilde. Halle, Nova Acta Leop., **44** (1883), 343-459.

SCHOUTE, P. H. Mehrdimensionale Geometrie. 2. Teil: Die Polytope. Leipzig, 1905.

STEINITZ, E. Polyeder und Raumeinteilungen. Encykl. math. Wiss. III, AB 12, 1916.

CHAPTER VIII

MENSURATION. CONTENT

1. THE simplest enclosed regions in space of any number of dimensions are comprised in the series: line-segment, square, cube, . . ., regular orthotope. The measure of length is a unit line-segment, the length of a straight line being measured by the number of times it contains the unit. The area or content of a rectangle is measured similarly by the number of times it contains the unit square, and so on. We need not enter in detail into the question of measuring a rectangular region when the unit is not contained an exact number of times. We assume that fractional parts of the unit can be measured, and the same principle applies without modification in n dimensions. In S_n the unit of content is an orthotope whose edges are each of unit length. The content of an orthotope with edges $a_1, \ldots, a_n$ is represented by the product $a_1a_2 \ldots a_n$.

A region whose boundaries are not rectangular cannot be divided into orthotopes without an unlimited division of the unit, and the process of integration has to be applied.

2. **The Prism.** Consider first a prism of the first species whose base is a polytope $(\mathrm{Po})_{n-1}$ of content C. If h is its height, the content of the prism is

$$V = Ch. \qquad (2{\cdot}1)$$

If the axis makes an angle θ with the normal to the base, and a is the length of the axis, the height $h = a \cos \theta$, and

$$V = Ca \cos \theta. \qquad (2{\cdot}2)$$

A prism of the second species is generated by a base-polytope $(\mathrm{Po})_{n-2}$ of content C moving first through a distance a_1 in a direction making an angle θ_1 with the normal to C which lies in the $(n-1)$-flat (Ca_1), thus generating a prism of the first species with content $Ca_1 \cos \theta_1$; then this prism, moving through a distance a_2 in a direction making an angle θ_2 with the normal to (Ca_1). The content of the prism is then

$$V = C\, a_1 a_2 \cos \theta_1 \cos \theta_2. \quad . \quad . \quad (2\cdot3)$$

Generally, the content of a prism of species r with base a polytope $(\text{Po})_{n-r}$ of content C_{n-r} and axes $a_1, \ldots, a_r$ is

$$V(C_{n-r}, a_1, \ldots, a_r) = C_{n-r} a_1 \ldots a_r \cos \theta_1 \ldots \cos \theta_r, \ldots \quad (2\cdot4)$$

where θ_s is the angle which the axis a_s makes with the normal to the $(n - r + s - 1)$-flat $(C_{n-r} a_1 \ldots a_{s-1})$ which lies in the $(n - r + s)$-flat $(C_{n-r} a_1 \ldots a_s)$.

If the angles θ_s are all right angles, each axis is normal to C_{n-r}, and the r-flat determined by the r axes drawn through a point O of C_{n-r} is completely orthogonal to C_{n-r}. Moreover in this case the axes themselves are mutually orthogonal, and determine an orthotope of content $a_1 \ldots a_r$. The content of the *right prism* of species r is then

$$C_{n-r} a_1 \ldots a_r, \quad . \quad . \quad . \quad (2\cdot5)$$

i.e. the product of the base and the content of the orthotope formed by the axes.

In the general case the axes form an oblique parallelotope. $a_s \cos \theta_s$ is the projection of a_s on the r-flat S_r which is completely orthogonal to C_{n-r}. The lines $a_s \cos \theta_s (s = 1, \ldots, r)$ form a parallelotope, which is the projection on S_r of the parallelotope formed by the axes. Hence *the content of a prism of any species is the product of the content of the base-polytope* $(\text{Po})_{n-r}$ *and that of the projection of the axes-parallelotope on the normal* S_r.

Hence if $\phi_1, \phi_2, \ldots, \phi_r$ are the angles which the r-flat containing the axes makes with the r-flat normal to the base, and A_r is the content of the parallelotope formed by the axes, C_{n-r} that of the base,

$$V(C_{n-r}, a_1, \ldots, a_r) = C_{n-r} A_r \cos \phi_1 \cos \phi_2 \ldots \cos \phi_r. \quad (2\cdot6)$$

This includes of course the parallelotope.

3. **The Parallelotope.** We shall obtain an expression for the content of a parallelotope in terms of the lengths of the edges a_r and the angles θ_{rs} between them. Let the content be denoted by $V(a_1, \ldots, a_n)$. Take a system of rectangular

co-ordinates with the origin at a vertex, and let the direction-cosines of the edge a_r be c^r_ν.* Let θ_r be the angle between the edge a_r and the normal to the $(r-1)$-flat $(a_1, \ldots, a_{r-1})$ which lies in the r-flat $(a_1, \ldots, a_r)$. Then

$$V(a_1, \ldots, a_r) = V(a_1, \ldots, a_{r-1})a_r \cos \theta_r.$$

Let l_ν be the direction-cosines of the normal. Then since it lies in $(a_1, \ldots, a_r)$

$$\rho l_\nu = \sum_{i=1}^{r} u_i c^i_\nu \qquad (\nu = 1, \ldots, n)$$

where $u_1, \ldots, u_r$ are r homogeneous parameters. Also since it is perpendicular to each of the axes $a_1, \ldots, a_{r-1}$,

$$\sum_{\nu=1}^{n} l_\nu c^j_\nu = 0 \qquad (j = 1, \ldots, r-1),$$

i.e.

$$\sum_{\nu=1}^{n} \sum_{i=1}^{r} u_i c^i_\nu c^j_\nu = 0 \qquad (j = 1, \ldots, r-1),$$

i.e.

$$\sum_{i=1}^{r} (u_i \sum_{\nu=1}^{n} c^i_\nu c^j_\nu) = 0.$$

But

$$\sum_{\nu=1}^{n} c^i_\nu c^j_\nu = \cos \theta_{ij} = c_{ij}, \text{ say,}$$

and

$$\sum_{\nu=1}^{n} c^i_\nu c^i_\nu = 1.$$

Hence

$$\sum_{i=1}^{r} u_i \cos \theta_{ij} = 0 \qquad (j = 1, \ldots, r-1).$$

These $r-1$ equations determine the ratios of $u_1, \ldots, u_r$. Let C_{ij} be the co-factor of c_{ij} in the determinant $| c_{rr} |$.

Then

$$u_i = C_{ir}.$$

* Throughout this and the next two paragraphs with a few obvious exceptions upper indices are used not for powers but as diacritical marks, as in the tensor-calculus.

Now $$\rho \cos \theta_r = \rho \sum_{\nu=1}^{n} l_\nu c_\nu^r = \sum_{\nu=1}^{n} (c_\nu^r \sum_{i=1}^{r} u_i c_\nu^i)$$

$$= \sum_{\nu=1}^{n} \sum_{i=1}^{r} C_{ir} c_\nu^r c_\nu^i = \sum_{i=1}^{r} (C_{ir} \sum_{\nu=1}^{n} c_\nu^r c_\nu^i)$$

$$= \sum_{i=1}^{r} C_{ir} c_{ir} = | c_{rr} |,$$

and $$\rho l_\nu = \sum_{i=1}^{r} C_{ir} c_\nu^i ;$$

hence $$\rho^2 = \sum_{i=1}^{r} \sum_{j=1}^{r} (C_{ir} C_{jr} \sum_{\nu=1}^{n} c_\nu^i c_\nu^j) = \sum_{i=1}^{r} \sum_{j=1}^{r} C_{ir} C_{jr} c_{ij}$$

$$= \sum_{i=1}^{r} (C_{ir} \sum_{j=1}^{r} C_{jr} c_{ij}).$$

But $$\sum_{j=1}^{r} C_{jr} c_{ij} = 0 \text{ if } i \neq r, \text{ and } = | c_{rr} | \text{ if } i = r.$$

Hence $$\rho^2 = C_{rr} | c_{rr} |.$$

We have then $$\cos^2 \theta_r = \frac{| c_{rr} |}{C_{rr}}.$$

The determinant $$| c_{rr} | = \begin{vmatrix} 1 & c_{12} & \cdots & c_{1r} \\ c_{21} & 1 & \cdots & c_{2r} \\ \cdot & \cdot & & \cdot \\ c_{r1} & c_{r2} & \cdots & 1 \end{vmatrix}$$

is a function of the angles θ_{ij} between the pairs of axes $a_1, \ldots, a_r$. For two axes a_1, a_2 it reduces to $1 - c_{12}^2 = \sin^2 \theta_{12}$. Its positive square root is therefore called the *sine of the r-dimensional angle* whose edges are $a_1, \ldots, a_r$, and is denoted by $\sin (a_1, \ldots, a_r)$. C_{rr} is the determinant of the same form for $r - 1$.

Hence

$$\cos \theta_r = \frac{\sin (a_1, \ldots, a_r)}{\sin (a_1, \ldots, a_{r-1})}.$$

Now

$$V(a_1, a_2) = a_1a_2 \sin\theta_{12} = a_1a_2 \sin(a_1, a_2),$$

$$V(a_1, a_2, a_3) = V(a_1, a_2)a_3 \cos\theta_3 = a_1a_2a_3 \sin(a_1, a_2) \frac{\sin(a_1, a_2, a_3)}{\sin(a_1, a_2)}$$

$$= a\, a_2a_3 \sin(a_1, a_2, a_3),$$

and generally

$$V(a_1, a_2, \ldots, a_n) = a_1a_2 \ldots a_n \sin(a_1, a_2, \ldots, a_n). \tag{3·1}$$

We have

$$\sin^2(a_1, \ldots, a_n) = \begin{vmatrix} 1 & c_{12} & \ldots & c_{1n} \\ \cdot & \cdot & & \cdot \\ c_{n1} & c_{n2} & \ldots & 1 \end{vmatrix}$$

$$= \begin{vmatrix} \Sigma c_\nu^1 c_\nu^1 & \Sigma c_\nu^1 c_\nu^2 & \ldots & \Sigma c_\nu^1 c_\nu^n \\ \cdot & \cdot & \cdot & \cdot \\ \Sigma c_\nu^n c_\nu^1 & \Sigma c_\nu^n c_\nu^2 & \ldots & \Sigma c_\nu^n c_\nu^n \end{vmatrix}$$

$$= \begin{vmatrix} c_1^1 & c_2^1 & \ldots & c_n^1 \\ \cdot & \cdot & & \cdot \\ c_1^n & c_2^n & \ldots & c_n^n \end{vmatrix}^2.$$

Hence

$$V(a_1, a_2, \ldots, a_n) = a_1a_2 \ldots a_n \,| c_n^n |. \tag{3·2}$$

Let x_ν^r be the co-ordinates of the end of the edge a_r through O, then $x_\nu^r = a_r c_\nu^r$, and

$$V(a_1, \ldots, a_n) = \begin{vmatrix} x_1^1 & x_2^1 & \ldots & x_n^1 \\ \cdot & \cdot & & \cdot \\ x_1^n & x_2^n & \ldots & x_n^n \end{vmatrix}. \tag{3·3}$$

We have similarly for a prism of species r, referred to rectangular axes, with the axes $x_{r+1}, \ldots, x_n$ in the base C_{n-r},

$$V(C_{n-r}, a_1, \ldots, a_r) = C_{n-r}a_1 \ldots a_r\sqrt{| c_r^{rr} |}$$
$$= C_{n-r}a_1 \ldots a_r| c_r^r |, \tag{3·4}$$

where

$$c_r^{\lambda\mu} = \sum_{\nu=1}^{r} c_\nu^\lambda c_\nu^\mu,$$

and

$$V(C_{n-r}, a_1, \ldots, a_r) = C_{n-r} \begin{vmatrix} x_1^1 & \ldots & x_r^1 \\ \cdot & \cdot & \cdot \\ x_1^r & \ldots & x_r^r \end{vmatrix}. \tag{3·5}$$

The determinant $|x_r^r|$ is therefore equal to the projection on the r-flat of $(x_1, \ldots, x_r)$ of the parallelotope whose edges are the lines joining the origin to the r points with co-ordinates (x_ν^r).

4. The Pyramid. Consider first a pyramid of the first species whose base is a polytope $(\text{Po})_{n-1}$ and vertex O. By a series of hyperplanes parallel to the base it is divided into slabs or thin prisms of thickness dx. The content of the section at a distance x from the vertex is proportional to the $(n-1)$th power of the distance. Hence if C is the content of the base and h the altitude, the content of the pyramid is

$$V = \int_0^h \frac{x^{n-1}}{h^{n-1}} C dx = \frac{1}{n} Ch, \qquad (4\cdot 1)$$

i.e. equal to one-nth of the content of a prism of first species with the same base and altitude.

Consider now a pyramid of species r whose base is a polytope $(\text{Po})_{n-r}$ of content C_{n-r}, and vertices $A_1, \ldots, A_r$. Let h_1 be the distance of A_1 from the base C_{n-r}, h_2 the distance of A_2 from the $(n-r+1)$-flat $(C_{n-r}A_1)$, and so on. Then

$$V = \frac{C_{n-r}h_1h_2 \ldots h_r}{(n-r+1)(n-r+2) \ldots n}.$$

Let O be any vertex of the base, $OA_s = a_s$ and θ_s the angle which OA_s makes with the normal to $(C_{n-r}A_1 \ldots A_{s-1})$ which lies in $(C_{n-r}A_1 \ldots A_s)$. Then

$$V = \frac{C_{n-r}a_1a_2 \ldots a_r}{n(n-1) \ldots (n-r+1)} \cos\theta_1 \cos\theta_2 \ldots \cos\theta_r. \qquad (4\cdot 2)$$

If rectangular axes are taken with origin O and the axes $x_{r+1}, \ldots, x_n$ lying in the base, and if (x_ν^s) are the co-ordinates of the vertex A_s

$$V = \frac{(n-r)!}{n!} C_{n-r} \begin{vmatrix} x_1^1 & \ldots & x_r^1 \\ \cdot & \cdot & \cdot \\ x_1^r & \ldots & x_r^r \end{vmatrix}. \qquad (4\cdot 3)$$

Hence also the content of a simplex with one vertex at the origin and n vertices A_s with rectangular co-ordinates (x_ν^s) is

$$V = \frac{1}{n!}\begin{vmatrix} x_1^1 & \cdots & x_n^1 \\ \cdot & \cdot & \cdot \\ x_1^n & \cdots & x_n^n \end{vmatrix}. \qquad (4{\cdot}4)$$

If the co-ordinates of the vertices $A_1, \ldots, A_{n+1}$ referred to any system of rectangular axes are (x_ν^r), and those referred to A_{n+1} are (y_ν^r), with $y_\nu^{n+1} = 0$, we have

$$y_\nu^r = x_\nu^r + x_\nu^{n+1},$$

and $$n!\,V = |y_n^n| = |x_n^n + x_n^{n+1}| = \begin{vmatrix} 1 & x_1^1 & \cdots & x_n^1 \\ \cdot & \cdot & & \cdot \\ 1 & x_1^n & \cdots & x_n^n \\ 1 & x_1^{n+1} & \cdots & x_n^{n+1} \end{vmatrix}. \qquad (4{\cdot}5)$$

Again, taking the origin at the vertex A_{n+1} let the direction-cosines of the edges $A_{n+1}A_r$ through A_{n+1} be c_ν^r. Then

$$n!\,V = a_1 \ldots a_n\,|c_n^n|.$$

Squaring the determinant and denoting by c_{rs} the cosine of the angle between the edges a_r, a_s we have

$$|c_n^n|^2 = \begin{vmatrix} 1 & c_{12} & \cdots & c_{1n} \\ \cdot & \cdot & & \cdot \\ c_{n1} & c_{n2} & \cdots & c_{nn} \end{vmatrix} = \sin^2(a_1, \ldots, a_n)$$

hence

$$n!\,V = a_1 \ldots a_n \sin(a_1, \ldots, a_n). \qquad (4{\cdot}6)$$

5. Content of a Simplex in Terms of the Lengths of the Edges. We may write the equation

$$n!\,V = \begin{vmatrix} x_1^1 & \cdots & x_n^1 & 1 \\ \cdot & \cdot & \cdot & \cdot \\ x_1^{n+1} & \cdots & x_n^{n+1} & 1 \end{vmatrix}$$

in the form

$$(-1)^n 2^n n!\,V = \begin{vmatrix} 1 & 0 & \cdots & 0 & 0 \\ \Sigma x_\nu^1 x_\nu^1 & -2x_1^1 & \cdots & -2x_n^1 & 1 \\ \cdot & \cdot & \cdot & \cdot & \cdot \\ \Sigma x_\nu^{n+1} x_\nu^{n+1} & -2x_1^{n+1} & \cdots & -2x_n^{n+1} & 1 \end{vmatrix}$$

or $$(-1)^{2n+1}n!\,V = \begin{vmatrix} 0 & 0 & \dots & 0 & 1 \\ 1 & x_1^1 & \dots & x_n^1 & \Sigma x_\nu^1 x_\nu^1 \\ \cdot & \cdot & \cdot & \cdot & \cdot \\ 1 & x_1^{n+1} & \dots & x_n^{n+1} & \Sigma x_\nu^{n+1} x_\nu^{n+1} \end{vmatrix}.$$

Multiply these together by rows. The elements of the first row and column in the product are all units except the first which is zero. Take any element of the diagonal

$$\Sigma x_\nu^r x_\nu^r - 2\Sigma x_\nu^r x_\nu^r + \Sigma x_\nu^r x_\nu^r = 0.$$

The element in the $(r+1)$th row and $(s+1)$th column is

$$\Sigma x_\nu^r x_\nu^r - 2\Sigma x_\nu^r x_\nu^s + \Sigma x_\nu^s x_\nu^s = \Sigma(x_\nu^r - x_\nu^s)^2 = a_{rs}^2,$$

where a_{rs} is the length of the edge A_rA_s.
Hence

$$(-1)^{n+1}2^n(n!)^2V^2 = \begin{vmatrix} 0 & 1 & 1 & \dots & 1 & 1 \\ 1 & 0 & a_{12}^2 & \dots & a_{1n}^2 & a_{1,n+1}^2 \\ 1 & a_{21}^2 & 0 & \dots & a_{2n}^2 & a_{2,n+1}^2 \\ \cdot & \cdot & \cdot & \cdot & \cdot & \cdot \\ 1 & a_{n+1,1}^2 & a_{n+1,2}^2 & \dots & a_{n+1,n}^2 & 0 \end{vmatrix}. \quad (5\cdot1)$$

For a regular simplex a_{rs} = constant = a, and the determinant becomes, on subtracting the last column from each of the others except the first,

$$\begin{vmatrix} 0 & 0 & 0 & \dots & 0 & 1 \\ 1 & -a^2 & 0 & \dots & 0 & a^2 \\ 1 & 0 & -a^2 & \dots & 0 & a^2 \\ \cdot & \cdot & \cdot & \cdot & \cdot & \cdot \\ 1 & 0 & 0 & \dots & -a^2 & a^2 \\ 1 & a^2 & a^2 & \dots & a^2 & 0 \end{vmatrix} = (-1)^{n+1} \begin{vmatrix} 1 & -a^2 & 0 & \dots & 0 \\ 1 & 0 & -a^2 & \dots & 0 \\ \cdot & \cdot & \cdot & \cdot & \cdot \\ 1 & 0 & 0 & \dots & -a^2 \\ 1 & a^2 & a^2 & \dots & a^2 \end{vmatrix}$$

$$= (-1)^{n+1}a^{2n} \begin{vmatrix} 1 & -1 & 0 & \dots & 0 \\ 1 & 0 & -1 & \dots & 0 \\ \cdot & \cdot & \cdot & \cdot & \cdot \\ 1 & 0 & 0 & \dots & -1 \\ n+1 & 0 & 0 & \dots & 0 \end{vmatrix} = (-1)^{n+1}(n+1)a^{2n}.$$

Hence the content of a regular simplex of edge a is

$$V = \frac{a^n}{n!}\sqrt{\frac{n+1}{2^n}}. \quad (5\cdot2)$$

This may also be obtained, without using determinants, by elementary geometry. Let p_r be the altitude A_rN_r of a regular simplex $S(r + 1)$. The foot of this altitude N_r is at the centroid of a face $S(n)$, and this centroid divides the altitude $A_{r-1}N_{r-1}$ so that $A_{r-1}N_{r-1} = r \,.\, N_rN_{r-1}$. We have then

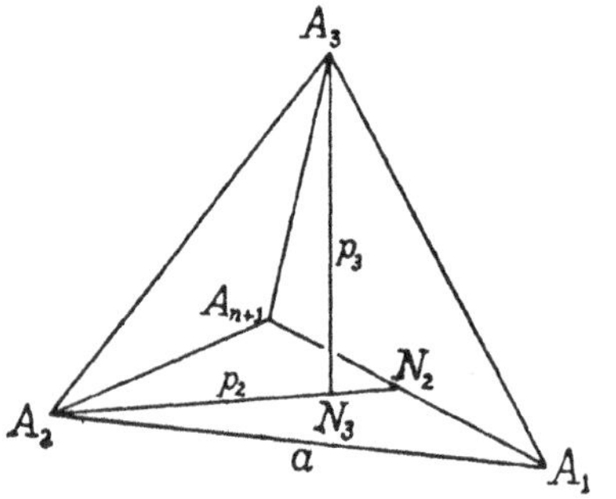

FIG. 35.

$$p_r^2 + \left(\frac{r-1}{r}\right)^2 p_{r-1}^2 = a^2.$$

Now

$$p_2^2 = a^2 - \frac{1}{4}a^2 = \frac{1}{2} \cdot \frac{3}{2}a^2,$$

$$p_3^2 = a^2 - \frac{4}{9}p_2^2 = \frac{1}{2} \cdot \frac{4}{3}a^2.$$

Assume that

$$p_{r-1}^2 = \frac{1}{2}\,\frac{r}{r-1}a^2,$$

then $$p_r^2 = a^2 - \left(\frac{r-1}{r}\right)^2 \frac{1}{2}\,\frac{r}{r-1}\,a^2 = \frac{1}{2}\,\frac{r+1}{r}a^2;$$

hence by induction this is true.

$$\text{Hence } V = \frac{1}{n!}\,ap_2p_3 \,.\,.\,.\, p_n$$

$$= \frac{1}{n!}\,a^n\sqrt{\left(\frac{1}{2^{n-1}}\,\frac{3}{2}\,\frac{4}{3}\,.\,.\,.\,\frac{n}{n-1}\,\frac{n+1}{n}\right)}$$

$$= \frac{a^n}{n!}\sqrt{\frac{n+1}{2^n}}.$$

6. Pyramid of Second Species. A section of a pyramid of the second species by a hyperplane which is parallel to the base α and also parallel to the vertex-edge is a polytope of $n - 1$ dimensions with two ends which are polytopes with edges parallel to corresponding edges of α, corresponding vertices of the ends being joined by edges parallel to the vertex-edge.

$PQRSA_1A_2$ (Fig. 36) represents a pyramid of the second species in S_4. A section parallel to the plane PQRS cuts the planes A_1PQ, A_1QR, etc., in lines parallel to PQ, QR, etc., so that the two pyramidal 3-dimensional boundaries A_1PQRS and

A_2PQRS are cut in two polygons $P_1Q_1R_1S_1$ and $P_2Q_2R_2S_2$ whose corresponding sides are parallel to the corresponding sides of PQRS. If also the hyperplane is parallel to A_1A_2 the edges P_1P_2, Q_1Q_2, . . . in which the hyperplane cuts the planes A_1A_2P, A_1A_2Q, etc., are all parallel to A_1A_2. Hence the section is a prism.

Let α be the area of the polygon PQRS, c the length of A_1A_2, and h the perpendicular distance between the line A_1A_2 and the plane PQRS, θ the angle between the direction of the

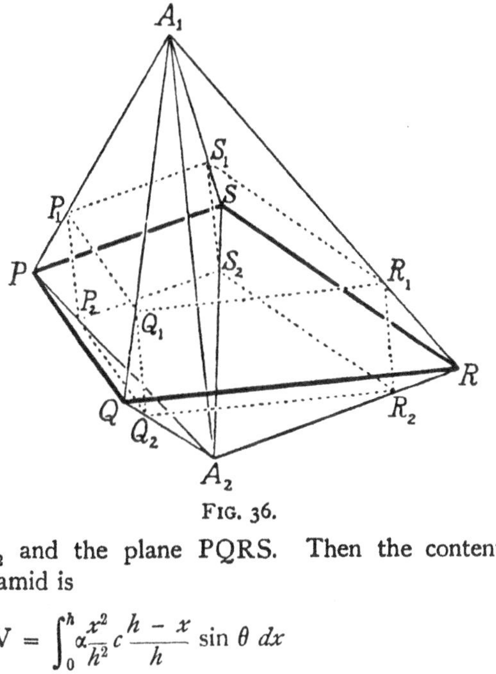

FIG. 36.

line A_1A_2 and the plane PQRS. Then the content of the hyperpyramid is

$$V = \int_0^h \alpha \frac{x^2}{h^2} c \frac{h - x}{h} \sin\theta \, dx$$

$$= \frac{\alpha c}{h^3} \sin\theta \left[\frac{1}{3} h x^3 - \frac{1}{4} x^4\right]_0^h = \frac{1}{12} \frac{\alpha c}{h^3} \sin\theta \, . \, h^4$$

$$= \frac{1}{12} \alpha c h \sin\theta. \quad . \quad . \quad . \quad . \quad . \quad . \quad (6{\cdot}1)$$

This is an extension to four dimensions of the well-known expression for the volume of a tetrahedron in terms of two opposite edges.

For a pyramid of second species in S_n we have similarly

$$V = \int_0^h \alpha \frac{x^{n-2}}{h^{n-2}} c \frac{h-x}{h} \sin\theta \, dx$$

$$= \frac{\alpha c}{h^{n-1}} \sin\theta \left(\frac{1}{n-1} - \frac{1}{n}\right) h^n$$

$$= \frac{\alpha c h \sin\theta}{n(n-1)} \quad . \quad . \quad . \quad . \quad . \quad (6 \cdot 2)$$

7. Consider now a pyramid of the third species in S_5, with a base-polygon of area α, and three vertices A_1, A_2, A_3.

Let the base-polygon be $\alpha \equiv PQR \ldots$ and have N sides. The pyramid is bounded by three pyramids of second species αA_2A_3, αA_3A_1, αA_1A_2, and N simplexes $PQA_1A_2A_3$, $QRA_1A_2A_3$, . . .

Divide the pyramid by 4-flats parallel to the two planes α and $A_1A_2A_3$. The section by a 4-flat is a figure bounded by three N-gonal prisms $P_2Q_2R_2 \ldots P_3Q_3R_3 \ldots$, $P_3Q_3R_3 \ldots P_1Q_1R_1 \ldots$, $P_1Q_1R_1 \ldots P_2Q_2R_2 \ldots$, and N triangular prisms $P_1P_2P_3Q_1Q_2Q_3$, $Q_1Q_2Q_3R_1R_2R_3$, . . . Let h be the distance between the planes α and $A_1A_2A_3$, and x the distance of the 4-flat from the plane $A_1A_2A_3$. Then $P_1Q_1 = P_2Q_2 = P_3Q_3 = \frac{x}{h} PQ$, and these three lines are parallel to PQ; similarly for Q_1R_1, etc. Also

$$P_1P_2 = Q_1Q_2 = \ldots = \frac{h-x}{h} A_1A_2,$$

and these N lines are all parallel to A_1A_2; similarly for P_2P_3, etc. Hence the three polygons $P_1Q_1R_1 \ldots$, $P_2Q_2R_2 \ldots$, $P_3Q_3R_3 \ldots$ are congruent, with corresponding sides parallel.

We have first to find the content of the 4-dimensional section. Dividing it by 3-flats parallel to that of the prism

$$P_2Q_2R_2 \ldots P_3Q_3R_3 \ldots,$$

the section is a prism $P_2'Q_2'R_2' \ldots P_3'Q_3'R_3' \ldots$ If k is the distance between the 3-flat of the prism $P_2Q_2 \ldots P_3Q_3 \ldots$ and the plane $P_1Q_1 \ldots$, and y the distance of the 3-flat of section from the plane $P_1Q_1 \ldots$, we have

$$P_2'Q_2' = P_3'Q_3' = P_2Q_2 = P_3Q_3,$$
$$P_2'P_3' = Q_2'Q_3' = \ldots = \frac{y}{k}P_2P_3.$$

The section is therefore a prism whose base

$$P_2'Q_2'R_2' \ldots = P_2Q_2R_2 \ldots = \frac{x^2}{h^2}\alpha,$$

and axis

$$P_2'P_3' = \frac{y}{k}P_2P_3 = \frac{y}{k}\frac{h-x}{h}A_2A_3.$$

Let θ be the angle which A_2A_3 makes with the normal to α which lies in the 3-flat parallel to α and A_2A_3. Then the volume of the prism

$$P_2'Q_2'R_2' \ldots P_3'Q_3'R_3' = (P_2Q_2R_2 \ldots)P_2'P_3' \cos\theta,$$

and the content of the 4-dimensional section is

$$\int_0^k (P_2Q_2R_2 \ldots)\,\frac{y}{k}(P_2P_3)\cos\theta\,dy$$
$$= \tfrac{1}{2}(P_2Q_2R_2 \ldots)k(P_2P_3)\cos\theta.$$

Now $k = P_1P_2 \cos\phi$, where ϕ is the angle which P_1P_2 makes with the line which is normal to both α and P_2P_3 and which lies in the 4-flat parallel to α, P_2P_3, and P_1P_2. Hence the content of the 4-dimensional section

$$= \frac{1}{2}\frac{x^2}{h^2}\alpha\left(\frac{h-x}{h}\right)^2(A_1A_2)(A_2A_3)\cos\theta\cos\phi.$$

Hence finally the content of the 5-dimensional pyramid

$$= \frac{1}{2}\frac{\alpha}{h^4}(A_1A_2)(A_2A_3)\cos\theta\cos\phi\int_0^h x^2(h-x)^2dx$$
$$= \frac{1}{60}\alpha(A_1A_2)(A_2A_3)h\cos\theta\cos\phi.$$

Now $\frac{1}{2}(A_1A_2)(A_2A_3)\cos\theta\cos\phi$ is the projection of the area of the triangle $A_1A_2A_3$ on the plane which is completely orthogonal to α. Hence if the area of this projection is denoted by A' we have

$$V = \frac{1}{30}\alpha A'h \quad . \quad . \quad . \quad (7\cdot1)$$

Let us extend this now for a pyramid of third species in S_n, whose base is a polytope $(Po)_{n-3}$ of content α and vertices A_1, A_2, A_3. We have

$$V = \tfrac{1}{2}\alpha(A_1A_2)(A_2A_3)\cos\theta\cos\phi\int_0^h\left(\frac{x}{h}\right)^{n-3}\left(\frac{h-x}{h}\right)^2 dx.$$

and hence

$$V = \frac{2\alpha A'h}{n(n-1)(n-2)}; \qquad (7\cdot 2)$$

and in general the content of a pyramid of species r in S_n is

$$V = \frac{(r-1)!\,\alpha A'h}{n(n-1)\ldots(n-r+1)} = \frac{(r-1)!\,(n-r)!\,\alpha A'h}{n!}, \qquad (7\cdot 3)$$

where α is the content of the base-polytope $(Po)_{n-r}$, A' the projection of the content of the simplex $A_1A_2 \ldots A_r$ on an $(r-1)$-flat completely orthogonal to the base, and h the distance between the base and the $(r-1)$-flat $A_1A_2 \ldots A_r$.

If A is the content of the simplex formed by the vertices, and ϕ_1, ϕ_2, ... the $(r-1)$ or $(n-r)$ angles (whichever number is the less) which its $(r-1)$-flat makes with the base-$(n-r)$-flat,

$$V = \frac{\alpha Ah}{r\,.\,{}_nC_r}\sin\phi_1\sin\phi_2\ldots \qquad (7\cdot 4)$$

8. The Simplotope. The simplotope of type $(p, n-p)$ is generated by moving a simplex $S(p+1)$ so that one vertex traverses every point of a simplex $S(n-p+1)$. Let the latter simplex be divided into elementary parallelotopes. Then when one vertex of the simplex $S(p+1)$ traverses this $(n-p)$-dimensional parallelotope it generates a prism of species $n-p$ whose base is $S(p+1)$ and axes the edges of the parallelotope. Hence if $\phi_1, \ldots, \phi_{n-p}$ are the angles which the $(n-p)$-flat of $S(n-p+1)$ makes with the normal $(n-p)$-flat to $S(p+1)$, and C_p, C_{n-p} denote the contents of the two simplexes, we have by summing the contents of the prisms, according to the result of § 2, for the content of the simplotope

$$V = C_pC_{n-p}\cos\phi_1\ldots\cos\phi_{n-p}, \qquad (8\cdot 1)$$

or if C'_{n-p} is the projection of the simplex $S(n - p + 1)$ on the normal $(n - p)$-flat to $S(p + 1)$,

$$V = C_p C'_{n-p}. \qquad (8\cdot 2)$$

9. Prismoidal Formulæ. It is well known that Simpson's formula $V = \frac{1}{6}h(A + B + 4M)$ for the content of a figure bounded by two parallel ends, A and B being the contents of the two ends, M that of the parallel mid-section, and h the distance between the two ends, holds for figures in both two and three dimensions for which the content of any section parallel to the ends is a *cubic* function of the distance from one end. Simpson's rule was applied only at first in the case of a quadratic function, and this is the case for a prismoid in three dimensions. For a prismoid in four dimensions it can be shown that the section is a cubic polynomial, and that Simpson's rule still holds. In higher dimensions polynomials of higher degree occur and the rule has to be extended.

10. Consider the general case of a figure bounded by two parallel ends, distant h apart, and such that the content of the parallel section at a distance λh from one end is a polynomial in λ of degree r. We shall call this a *prismoidal figure of species r*. Except for the two flat ends the boundaries need not necessarily be flat, but might be curved, e.g. a quadric surface bounded by two planes parallel to principal planes.

The content of the section may be expressed in the form

$$M(\lambda) = a_0(1-\lambda)^r + a_1(1-\lambda)^{r-1}\lambda + \ldots + a_{r-1}(1-\lambda)\lambda^{r-1} + a_r\lambda^r.$$

Then the whole content is

$$V = \int_0^1 M(\lambda) h d\lambda.$$

Now

$$\int_0^1 \lambda^p(1 - \lambda)^{r-p} d\lambda = \frac{1}{(r+1)\,.\,{}_rC_p}.$$

Hence

$$\frac{r+1}{h}V = a_0 + \frac{a_1}{{}_rC_1} + \frac{a_2}{{}_rC_2} + \ldots + \frac{a_{r-1}}{{}_rC_1} + a_r$$

$$= (a_0 + a_r) + (a_1 + a_{r-1})/{}_rC_1 + \ldots,$$

the last term being

$$\{a_{\frac{1}{2}(r-1)} + a_{\frac{1}{2}(r+1)}\}/{}_rC_{\frac{1}{2}(r-1)} \text{ if } r \text{ is odd}$$

and

$$a_{\frac{1}{2}r}/{}_rC_{\frac{1}{2}r} \text{ if } r \text{ is even.}$$

Now take two sections equidistant from the two ends. We have

$$\begin{aligned} M(\lambda) + M(1 - \lambda) = {} & (a_0 + a_r)\{\lambda^r + (1 - \lambda)^r\} \\ & + (a_1 + a_{r-1})\lambda(1 - \lambda)\{\lambda^{r-2} + (1 - \lambda)^{r-2}\} + \ldots, \end{aligned}$$

the last term being

$$\{a_{\frac{1}{2}(r-1)} + a_{\frac{1}{2}(r+1)}\}\{\lambda(1 - \lambda)\}^{\frac{1}{2}(r-1)} \text{ if } r \text{ is odd}$$

and

$$2a_{\frac{1}{2}r}\{\lambda(1 - \lambda)\}^{\frac{1}{2}r} \text{ if } r \text{ is even.}$$

V is thus expressed in terms of the $[\frac{1}{2}r] + 1$ quantities $(a_p + a_{r-p})$; and these can be expressed linearly in terms of the same number of quantities $M(\lambda) + M(1 - \lambda)$. Whether $r = 2k$ or $2k + 1$ the number of pairs of parallel sections required is the same, viz. $k + 1$. Thus if

$$T_{s, r} = M\left(\frac{s}{r}\right) + M\left(1 - \frac{s}{r}\right),$$

V can be expressed in terms of the $k + 1$ quantities $T_{s, 2k+1}$ or $T_{s, 2k}$ $(s = 0, 1, \ldots, k)$. When the denominator is $2k$, $T_{k, 2k} = 2M(\frac{1}{2})$, i.e. double the mid-section. It is clear that any formula of this type which holds for a particular value of r will hold for any smaller value of r, since a polynomial of degree $r - 1$ can be regarded as a polynomial of degree r in which the highest coefficient is zero.

11. We shall find it more convenient now to express $M(\lambda)$ in the form

$$M(\lambda) = c_0 + c_1\lambda + c_2\lambda^2 + \ldots + c_r\lambda^r, \quad . \quad (11\cdot1)$$

Then

$$V/h = \int_0^1 M(\lambda)d\lambda = c_0 + \tfrac{1}{2}c_1 + \tfrac{1}{3}c_2 + \ldots + \frac{1}{r+1}c_r, \quad (11\cdot2)$$

and

$$\begin{aligned} M(\lambda) + M(1 - \lambda) = {} & 2c_0 + c_1 + c_2\{\lambda^2 + (1 - \lambda)^2\} \\ & + \ldots + c_r\{\lambda^r + (1 - \lambda)^r\}. \quad (11\cdot3) \end{aligned}$$

Write

$$\lambda^r + (1 - \lambda)^r = f_r \quad . \quad . \quad . \quad (11\cdot4)$$

and

$$\lambda(1 - \lambda) = y. \quad . \quad . \quad . \quad (11\cdot5)$$

If we require to express the value of λ in f_r and y we shall use the notation $f_r(s)$ and y_s for f_r and y when $\lambda = s/(2k)$ or $s/(2k + 1)$.

Then we have

$$f_p - yf_{p-1} = \{\lambda^p + (1 - \lambda)^p\} - \lambda(1 - \lambda)\{\lambda^{p-1} + (1 - \lambda)^{p-1}\}$$
$$= \lambda^{p+1} + (1 - \lambda)^{p+1} = f_{p+1}. \quad . \quad (11\cdot6)$$

Now let

$$V/h = b_0T_0 + b_1T_1 + \ldots + b_kT_k,$$

where T_k is written for either $T_{k,\,2k+1}$ or $T_{k,\,2k}$.

Substituting from (11·3), viz.:

$$T_s = c_0f_0(s) + c_1f_1(s) + \ldots + c_rf_r(s)$$

we have

$$V/h = c_0\,\Sigma\,b_sf_0(s) + c_1\,\Sigma\,b_sf_1(s) + \ldots + c_r\,\Sigma\,b_sf_r(s),$$

summations being from $s = 0$ to $s = k$. Comparing this with (11·2), which is true for all values of r up to $2k + 1$, we have

$$\Sigma b_sf_i(s) = \frac{1}{i + 1}.$$

Since $f_1(s) = 1$ this gives

$$\sum_{s=0}^{k} b_s = \tfrac{1}{2}. \quad . \quad . \quad . \quad (11\cdot7)$$

Again by (11·6)

$$\Sigma b_sf_{i+2}(s) = \Sigma b_sf_{i+1}(s) - \Sigma b_sy_sf_i(s),$$

hence

$$\Sigma\,b_sy_sf_i(s) = \frac{1}{i + 2} - \frac{1}{i + 3} = \frac{1}{(i + 2)(i + 3)} \quad (i \gtrless r - 2)$$

and putting $i = 1$

$$\sum_{s=1}^{k} b_sy_s = \frac{1}{3\cdot4},$$

since $y_0 = 0$.

Again

$$\Sigma b_sy_sf_{i+2}(s) = \Sigma b_sy_sf_{i+1}(s) - \Sigma b_sy_s^2\,f_i(s),$$

hence

$$\Sigma b_s y_s^2 f_i(s) = \frac{1}{(i+3)(i+4)} - \frac{1}{(i+4)(i+5)}$$
$$= \frac{2}{(i+3)(i+4)(i+5)} \quad (i \gtrless r-4),$$

and putting $i = 1$

$$\sum_{s=1}^{k} b_s y_s^2 = \frac{2}{4 \cdot 5 \cdot 6}.$$

Proceeding in this way we can prove by induction that

$$\sum_{s=1}^{k} b_s y_s^p f_i(s) = \frac{p\,!}{(p+1+i)\ \ldots\ (2p+1+i)} \qquad (i \gtrless r-2p),$$

and putting $i = 1$

$$\sum_{s=1}^{k} b_s y_s^p = \frac{p\,!\,(p+1)\,!}{(2p+2)\,!} \quad (p \gtrless \tfrac{1}{2}r). \quad . \quad (11\cdot8)$$

Hence for all values of r up to $2k + 1$ we have two expressions for V, viz.

$$\left.\begin{aligned} V/h &= b_0 T_{0,2k} + b_1 T_{1,2k} + \ldots + b_k T_{k,2k} \\ \text{and}\quad V/h &= b_0' T_{0,2k+1} + b_1' T_{1,2k+1} + \ldots + b_k' T_{k,2k+1}, \end{aligned}\right\} \quad (11\cdot9)$$

where the coefficients are determined by the equations (11·7) and (11·8), λ being equal to $s/(2k)$ or $s/(2k+1)$ respectively for the two formulæ.

The following table gives the values of the coefficients in the formula

$$RV/h = B_0 T_0 + B_1 T_{1,m} + B_2 T_{2,m} + \ldots$$

up to $m = 7$:

m.	$r \gtrless$	B_0.	B_1.	B_2.	B_3.	R.
2	3	1	2			6
3	3	1	3			8
4	5	7	32	6		90
5	5	19	75	50		288
6	7	41	216	27	136	840
7	7	751	3577	1323	2989	17280

T_0 is the sum of the two ends, $T_{1,2} = T_{2,4} = T_{3,6} =$ double the middle section.

12. We shall now show that *a prismoid in S_n is a prismoidal figure of species $n - 1$.* In addition to the flat ends A, B, polytopes of $n - 1$ dimensions, the boundaries are pyramidoids of $n - 1$ dimensions. The prismoid may be divided into simpler figures in the following way. Joining a vertex of A to all the vertices of B we obtain a pyramid of the first species. The remaining figure can be divided into prismoids which have as bases a boundary of B and either A or a boundary of A. These may be divided further into prismoids having as bases boundaries of A and B of lower dimensions, giving ultimately pyramidoids of n dimensions, each with a $(p - 1)$-boundary of A and an $(n - p)$-boundary of B as bases.

Now consider a section of the prismoid parallel to the ends and at a distance λh from A; and in particular the section of the pyramidoid which has as bases the $(p - 1)$-boundary a of A and the $(n - p)$-boundary b of B. This section is a polytope generated by moving a polytope a' always parallel to itself so that one of its vertices moves over the whole boundary of a polytope b'; a' being a polytope similar to a but of content $(1 - \lambda)^{p-1}a$, and b' a polytope similar to b but of content $\lambda^{n-p}b$. The content of this section is thus proportional to the product $(1 - \lambda)^{p-1}\lambda^{n-p}ab$. Hence the content of the section of each region of the prismoid is represented by a polynomial of degree $n - 1$.

13. The Hypersphere. The content of a hypersphere of n dimensions, of radius a, is proportional to a^n, and we can write

$$V_n = K_n a^n,$$

where K_n is some numerical function of n which is to be determined. Dividing the hypersphere into thin slabs by a series of parallel hyperplanes, since each section is a hypersphere of $n - 1$ dimensions, we have

$$V_n = K_n a^n = 2\int_0^a K_{n-1}(a^2 - x^2)^{\frac{n-1}{2}} dx.$$

Put $x = a \cos\theta$, $dx = -a \sin\theta d\theta$. Then

$$K_n = 2K_{n-1}\int_0^{\frac{1}{2}\pi} \sin^n \theta \, d\theta$$

$$= 2K_{n-1}\frac{\Gamma\left(\frac{n+1}{2}\right)\Gamma\left(\frac{1}{2}\right)}{2\Gamma\left(\frac{n}{2}+1\right)},$$

i.e.

$$K_n = K_{n-1}\frac{\Gamma\left(\frac{n+1}{2}\right)}{\Gamma\left(\frac{n+2}{2}\right)}\pi^{\frac{1}{2}}.$$

Also $V_2 = \pi a^2$, therefore $K_2 = \pi$. Hence we find

$$K_n = \frac{\pi^{\frac{1}{2}n}}{\Gamma(\frac{1}{2}n + 1)};$$

and therefore

$$V_n = \frac{\pi^{\frac{1}{2}n}a^n}{\Gamma(\frac{1}{2}n + 1)}.$$

When n is even, $= 2m$,

$$V_{2m} = \frac{\pi^m a^{2m}}{m!};$$

and when n is odd, $= 2m + 1$,

$$V_{2m+1} = \frac{\pi^m m!\,(2a)^{2m+1}}{(2m+1)!}.$$

14. *Surface-content of a Hypersphere.* If the interior of a hypersphere is divided into concentric shells we have another method of finding the volume-content by integration. Let $S_n = C_n a^{n-1}$ be the surface-content of a hypersphere of n dimensions. Then

$$V_n = K_n a^n = \int_0^a C_n r^{n-1} dr = \frac{1}{n}C_n a^n.$$

Hence $C_n = nK_n$, and the surface-content is

$$S_n = \frac{2\pi^{\frac{1}{2}n}a^{n-1}}{\Gamma(\frac{1}{2}n)}.$$

When n is even, $= 2m$,

$$S_{2m} = \frac{2\pi^m a^{2m-1}}{(m-1)!};$$

when n is odd, $= 2m + 1$,

$$S_{2m+1} = \frac{2\pi^m m!\,(2a)^{2m}}{(2m)!}.$$

The following short table gives the volume and surface-contents of a hypersphere in space of n dimensions, $n = 2$ to 7:

n.	2.	3.	4.	5.	6.	7.
V_n	πa^2	$\frac{4}{3}\pi a^3$	$\frac{1}{2}\pi^2 a^4$	$\frac{8}{15}\pi^2 a^5$	$\frac{1}{6}\pi^3 a^6$	$\frac{16}{105}\pi^3 a^7$
S_n	$2\pi a$	$4\pi a^2$	$2\pi^2 a^3$	$\frac{8}{3}\pi^2 a^4$	$\pi^3 a^5$	$\frac{16}{15}\pi^3 a^6$

Ex. Find the centroid of (i) the volume-content, (ii) the surface-content of a hyperhemisphere in S_n.

(Ans. Distance from centre,

$$\text{(i)}\ \frac{\Gamma\frac{1}{2}(n+2)}{\sqrt{\pi}\,\Gamma\frac{1}{2}(n+3)}a, \qquad \text{(ii)}\ \frac{\Gamma\frac{1}{2}(n)}{\sqrt{\pi}\,\Gamma\frac{1}{2}(n+1)}a.)$$

15. Varieties of Revolution. If a hyperplane has $n - 1$ points fixed it has still one degree of freedom and can rotate about the $(n - 2)$-flat, which is determined by the $n - 1$ points, as axis. A V_{n-2} lying in the $(n - 1)$-flat will generate a V_{n-1} of revolution; every point of it describes a circle whose centre is the foot of the perpendicular to the axis from the given point, and any section by a plane perpendicular to the axis is a circle. Let the equation of the generating V_{n-2} referred to $n - 2$ rectangular axes in the S_{n-2} axis of revolution, say $x_1, \ldots, x_{n-2}$, and one other axis, x_{n-1}, be

$$x_{n-1}^2 = f(x_1, \ldots, x_{n-2}).$$

Then the equation of the V_{n-1} of revolution is

$$x_{n-1}^2 + x_n^2 = f(x_1, \ldots, x_{n-2}).$$

This is called a *variety of revolution of the first species.*

If we take next $n - 2$ fixed points, the hyperplane has two degrees of freedom, and every point will generate a sphere. A

V_{n-3} lying in the hyperplane will generate a V_{n-1} *of revolution of the second species*, whose equation will be of the form

$$x_{n-2}^2 + x_{n-1}^2 + x_n^2 = f(x_1, \ldots, x_{n-3}).$$

Generally, a V_{n-1} of revolution of species p is generated by a V_{n-p-1} rotating about an $(n-p-1)$-flat lying in its $(n-p)$-flat; every point generates a hypersphere of $p+1$ dimensions. The equation of the variety of revolution is of the form

$$\sum_{n-p}^{n} x_\nu^2 = f(x_1, \ldots, x_{n-p-1}).$$

A hypersphere of n dimensions is a variety of revolution of species $n-1$. If the generating variety is an $(n-p-1)$-flat the variety of revolution of species p is a quadric hypercone of species $n-p-1$. A quadric hypercone (of revolution) of species r is a variety of revolution of species $n-r-1$.

16. *Content of a variety of revolution of species $n-2$, generated by a curve*

$$x_{n-1}^2 = f(x_n)$$

rotating about the axis of x_n.

Dividing the content by hyperplanes perpendicular to the axis, each section is a hypersphere and the total content is

$$\frac{\pi^{\frac{1}{2}(n-1)}}{\Gamma\left(\frac{n+1}{2}\right)}\int_a^\beta x_{n-1}^{n-1}dx_n = \frac{\pi^{\frac{1}{2}(n-1)}}{\Gamma\left(\frac{n+1}{2}\right)}\int_a^\beta \{f(x)\}^{\frac{1}{2}(n-1)}dx.$$

The surface is similarly

$$\frac{(n-1)\pi^{\frac{1}{2}(n-1)}}{\Gamma\left(\frac{n+1}{2}\right)}\int_a^\beta \{f(x)\}^{\frac{1}{2}(n-2)}dx.$$

For a variety of revolution of species $p-1$ we have multiple integrals:

$$\text{Volume} = \frac{\pi^{\frac{1}{2}p}}{\Gamma(\frac{1}{2}p+1)}\int\ldots\int x_n^p dx_1 \ldots dx_{n-p},$$

$$\text{Surface} = \frac{p\pi^{\frac{1}{2}p}}{\Gamma(\frac{1}{2}p+1)}\int x_{n-p+1}^{p-1} dS_{n-p},$$

where dS_{n-p} is the element of surface of the generating variety.

17. Extensions of Pappus' Theorem. Consider first a variety of revolution of the first species.

$$V = \pi\int \dots \int x_{n-1}^2 dx_1 \dots dx_{n-2}.$$

Let $\bar{x}_{n-1}$ be the distance of the centroid of the generating figure from the axis of rotation, then

$$\bar{x}_{n-1}\int \dots \int dx_1 \dots dx_{n-1} = \int \dots \int x_{n-1} dx_1 \dots dx_{n-1}.$$

Let A be the content of the generating figure, then

$$A\bar{x}_{n-1} = \int \dots \int x_{n-1} dx_1 \dots dx_{n-1} = \tfrac{1}{2}\int \dots \int x_{n-1}^2 dx_1 \dots dx_{n-2}.$$

Hence $$2\pi\bar{x}_{n-1}A = V.$$

Again, let S be the surface-content of the generating figure, $\bar{x}_{n-1}$ the distance of the centroid of the surface from the axis of rotation, then

$$S\bar{x}_{n-1} = \int x_{n-1} dS.$$

Hence $$2\pi\bar{x}_{n-1}S = 2\pi\int x_{n-1} dS$$

= the surface-content of the figure of revolution,

i.e. *the volume-content of a variety of revolution of the first species is equal to the product of the volume-content of the generating figure and the length of the path described by the centroid of this volume*, and

The surface-content of a variety of revolution of the first species is equal to the product of the surface-content of the generating figure and the length of the path described by the centroid of this surface.

These results can be extended to varieties of revolution of any species $p - 1$. Let

$$\bar{x}_\nu^{p-1} = \frac{\int \dots \int x_\nu^{p-1} dx_1 \dots dx_{n-p} dx_\nu}{\int \dots \int dx_1 \dots dx_{n-p} dx_\nu}.$$

Then
$$V = \frac{\pi^{\frac{1}{2}p}}{\Gamma(\frac{1}{2}p + 1)}\int \cdots \int x_\nu^p dx_1 \ldots dx_{n-p}$$
$$= \frac{p\pi^{\frac{1}{2}p}}{\Gamma(\frac{1}{2}p + 1)}\int \cdots \int x_\nu^{p-1} dx_1 \ldots dx_{n-p} dx_\nu$$
$$= \frac{2\pi^{\frac{1}{2}p}}{\Gamma(\frac{1}{2}p)} \bar{x}_\nu^{p-1} A_{n-p+1},$$

where A_{n-p+1} is the volume-content of the generating figure.

Hence *the volume-content of a variety of revolution of species $p - 1$ is equal to the product of the volume-content of the generating figure and the surface-content of the hypersphere traced by the $(p - 1)$th centroid of the generating figure.*

Again if

$$\bar{x}_{n-p+1}^{p-1} = \frac{\int x_{n-p+1}^{p-1} dS_{n-p}}{\int dS_{n-p}},$$

then

$$S = \frac{p\pi^{\frac{1}{2}p}}{\Gamma(\frac{1}{2}p + 1)}\int x_{n-p+1}^{p-1} dS_{n-p} = \frac{2\pi^{\frac{1}{2}p}}{\Gamma(\frac{1}{2}p)} \bar{x}_{n-p+1}^{p-1} S_{n-p},$$

i.e. *the surface-content of a variety of revolution of species $p - 1$ is equal to the product of the surface-content of the generating figure and the surface-content of the hypersphere traced by the $(p - 1)$th centroid of this surface.*

REFERENCES

COTES, ROGER. Harmonia mensurarum sive Analysis et synthesis per rationum et angulorum mensuras promotae. Cambridge, 1722.

SCHLÄFLI, L. Theorie der vielfachen Kontinuität.
N. Denkschr. Schweiz. Ges. Natw., **38** (1901). (Published posthumously, written in 1850-2. Part transl. by Cayley, Q.J. Math., **2** (1858), 269-301 ; **3** (1860), 54-68, 97-108.)

SIMPSON, THOMAS. Of the areas of curves, etc., by approximation. Math. Dissertations, pp. 109-119. 1743.

CHAPTER IX

EULER'S THEOREM

1. Euler's Polyhedral Formula. If F, E, V denote the number of faces, edges, and vertices of a polyhedron, then, with certain limitations which are satisfied for "ordinary" types of polyhedra,

$$F - E + V = 2.$$

This famous theorem was published by Euler in 1752/3. Proofs of it are given in elementary text-books on solid geometry, but it will be well, before considering its extension to n dimensions, to consider it in some detail for three dimensions. Several distinct types of proof exist, and the relations between these throw much light on the theory of polyhedra and polytopes.

I. We shall begin with the commonest proof in which the polyhedron is built up, adding one face after another, until it is complete. This is similar to Euler's own proof, which, however, proceeds by dissection instead of building up.

Write $$F - E + V - 1 = \phi.$$

We assume that it is possible always to add new faces in such a way that every new face has only *consecutive* edges in contact with old faces. For the first face, an n-gon, say, we have $F = 1$, $E = n$, $V = n$, and therefore $\phi = 0$. When a second face is added, say an m-gon having one edge common to the n-gon, we have $F = 2$, $E = n + m - 1$, $V = n + m - 2$, and ϕ again has the value 0.

We can now prove by induction that, so long as the polyhedron is incomplete, $\phi = 0$. Assuming this for any stage of construction let an n-gon be added having m consecutive edges, and therefore $m + 1$ consecutive vertices, in contact with the old ones. Then, unaccented letters referring to the one stage and accented letters to the next, we have $F' = F + 1$, $E' = E + (n - m)$, $V' = V + (n - m - 1)$, and therefore $\phi' = F' - E' + V' - 1 = F - E + V - 1 = \phi = 0$.

When the last face is added, however, no new edges or vertices are added; $F' = F + 1$, $E' = E$, $V' = V$, and $\phi' = \phi + 1 = 1$.

Before examining this further we give three other alternative proofs, of which the first two are valid only for convex polyhedra.

II. *Lhuilier and Steiner's Proof.* Project the polyhedron orthogonally and we obtain a polygon covered twice over with a network of polygons (Fig. 37). This is always the case for a convex polyhedron since no line can meet its boundary in more than two points; it *may* be possible to project a non-convex polyhedron in this way by suitable choice of the plane of projection.

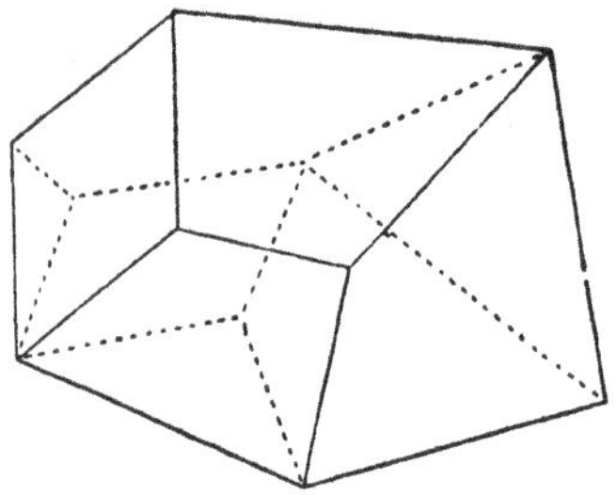

FIG. 37.

Let S be the sum of the angles of these polygons, and let the numbers of sides of the F faces be $n_1, n_2, \ldots, n_F$. Then

$$S = \Sigma(n_i - 2) = (2E - 2F)\pi,$$

since $\Sigma n_i = 2E$. But if the boundary polygon has n vertices, there are $V - n$ interior vertices, and we have also

$$S = 2(n - 2)\pi + (V - n)2\pi = (2V - 4)\pi.$$

Hence $\quad 2E - 2F = 2V - 4$, i.e. $F - E + V = 2$.

III. *Legendre's Proof.* Project the polyhedron from an interior point on to the surface of a sphere of unit radius with this point as centre. Since for a convex polyhedron every ray from an interior point cuts the boundary once only, the sphere will be covered completely, once over, by a network of spherical polygons. The area of a spherical triangle with angles α, β, γ is $(\alpha + \beta + \gamma - \pi)$, and that of a spherical polygon of n sides is $\Sigma\alpha - (n - 2)\pi$. Hence the total area of the spherical polygons is

$$4\pi = \Sigma\alpha - \Sigma(n_i - 2)\pi.$$

But $\Sigma\alpha = 2\pi V$, hence again we get the same result.

IV. *Von Staudt's Proof.* This proof is of quite a different character from the others, and does not admit of extension to n dimensions. We note in the first place that the edges of a polyhedron are joins both of adjacent vertices and of adjacent faces. Suppose now a path, starting with any chosen vertex, is drawn along the edges, leading to every vertex (see Fig. 38). The path may branch anywhere, but must not include any closed circuits. The number of edges included in this path is clearly V − 1. Now let another similar path be constructed, starting at one face and crossing the other edges, leading to every face. That it is possible to pass from any one face to any other face in this way follows since no face or group of faces is ring-fenced by the edges of the first path. Also there is only one path from one face to another, for if there were two, this double path would isolate a group of vertices and render the first path impossible. Hence the first path being chosen, the second is uniquely determined and crosses each of the edges left out of the first path. But the number of edges in the second path is F − 1. Hence

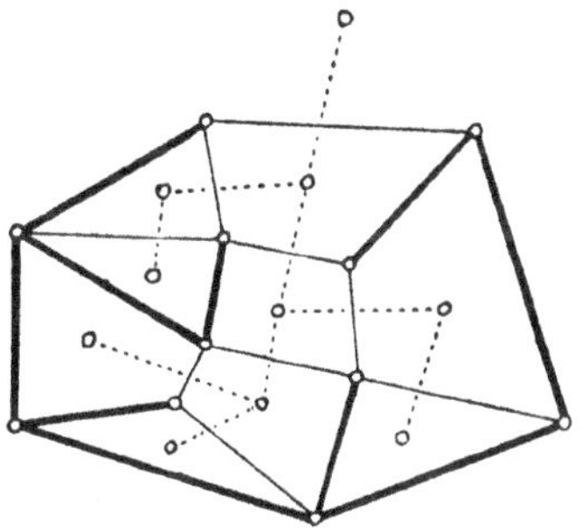

FIG. 38.

$$(V - 1) + (F - 1) = E, \text{ i.e. } F - E + V = 2.$$

The assumption made here, which we shall find to be an important one, is that any closed path consisting of edges of the polyhedron divides the polyhedron into two separate parts.

2. Steiner's and Legendre's proofs show that Euler's theorem is true for any convex polyhedron, but convexity is to a certain extent accidental, and a convex polyhedron might be transformed, for example by a dent or by pushing in one or more of the vertices, into a non-convex polyhedron with the same configurational numbers. Euler's relation corresponds to something more fundamental than convexity. Let us examine the condition which we assumed in the first proof, and the possibility of types of polyhedra which will necessarily violate this condition, and for which Euler's equation may not be true.

3. Connectivity of Polygons. We shall examine first the nature of the faces. We may dismiss the examination of the edges with the assumption that each edge is a simple line joining two separate points (consecutive vertices). A property of a convex polygon, which is true also for certain polygons which are not convex, is that any two interior * points can be connected by a path (curve or broken line) which is made up entirely of interior points. Such a polygon, or more strictly the interior region of such a polygon, is said to be *connected.* If, further, a simple line (not intersecting itself), joining any two points on the boundary and composed for the rest of interior points, divides the polygon into two separate regions, the polygon is said to be *simply-connected.* An example of a polygon which is not simply-connected is a ring-shaped polygon whose boundary consists of two separate polygons, one entirely within the other; such a polygon is doubly-connected. If there are several inner polygons as part of the boundary the region is multiply-connected.

4. Polyhedra with Ring-shaped Faces. Now clearly for a polyhedron which has one multiply-connected face the condition in Euler's proof cannot be satisfied, since this face has edges which are not consecutive. An example of such a polyhedron is one which is composed of two convex polyhedra having in common the hollow of the ring-shaped face (Fig. 39). Separating these, and supplying the missing face to one of them, we form two convex polyhedra, for which to-

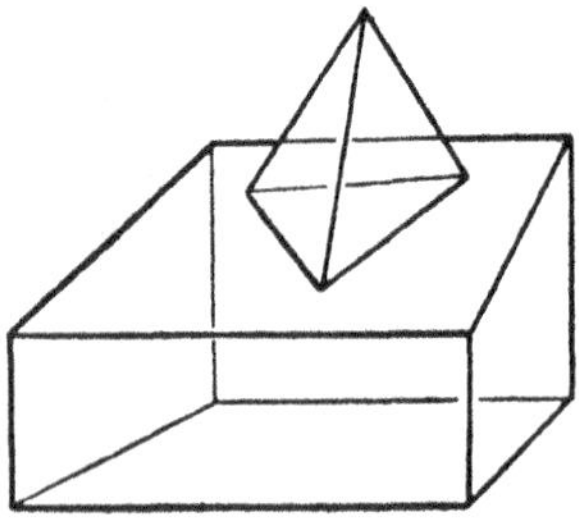

FIG. 39.

* An interior point is characterised by the property that every ray from it meets the boundary in an odd number of points.

gether $F - E + V = 4$. Hence for the original polyhedron $F - E + V = 3$.

If the face has m holes in it, each connected with a separate convex polyhedron, we have $F - E + V = 2(m + 1) - m = 2 + m$.

5. Connectivity of Polyhedra. A region in general, of any number of dimensions, is said to be connected when any two points of the region can be connected by a path which consists entirely of points belonging to the region. A region of three dimensions is said to be simply-connected when it is divided into separate regions by every partition, consisting of a simply-connected region of two dimensions, whose boundary is a simple closed curve lying entirely in the boundary of the three-dimensional region. The interior of an anchor-ring is an example of a three-dimensional region which is not simply-connected.

6. Ring-shaped Polyhedra. An example of a ring-shaped polyhedron is a picture-frame bevelled on both sides (Fig. 40).

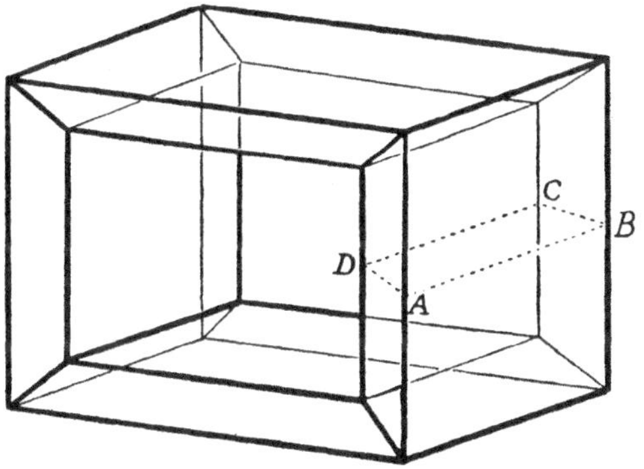

FIG. 40.

If this is divided by a partition such as ABCD, and two faces, with edges and vertices complete, are supplied to each side of the cut, it becomes simply-connected. Now, accents referring to the transformed polyhedron, we have

$$F' = F + 2,\ E' = E + 2n,\ V' = V + 2n.$$

Hence $$F - E + V = F' - E' + V' - 2 = 0,$$

and $\phi = -1$. If there are q such rings $\phi = 1 - 2q$.

7. Polyhedra with Cavities. The boundary of an ordinary polyhedron is a simply-connected region of two dimensions; that of a ring-shaped polyhedron is connected, but not simply connected. We may further have a polyhedron whose boundary is not even connected, but consists of distinct parts. An example of this is a polyhedron with a cavity inside, having an outer and an inner boundary. It is therefore equivalent to two separate polyhedra, for each of which $F - E + V$ has the value 2. Hence for the whole polyhedron $F - E + V = 4$, and $\phi = 3$. If there are r separate cavities,

$$F - E + V = 2(r + 1).$$

8. Eulerian Polyhedra. The three cases just considered can be summed up in one statement. For a polyhedron with p ring-shaped faces, q rings, and r cavities

$$F - E + V = 2 + p - 2q + 2r.$$

A simply-connected polyhedron whose boundary is simply-connected is called an Eulerian polyhedron, sometimes a spheroidal (kugelartig) polyhedron, and for such polyhedra the relation $F - E + V = 2$ is verified. The converse is not true, for a polyhedron may have a combination of singularities for which $p - 2q + 2r = 0$; for example, a ring-shaped polyhedron with two ring-shaped faces, a polyhedron with a ring-shaped cavity, and so on.

9. Incomplete Polyhedra and Polytopes. We have now to extend Euler's theorem to n dimensions, and we shall adapt the first form of proof, where we consider the polytope in process of construction. We have therefore to consider it both when complete and when in an incomplete state. When a simply-connected polyhedron is cut in two along a simply-connected polygon (in general skew) the two parts are incomplete polyhedra, each with one free rim. We may suppose that one of the parts takes with it the edges and vertices of the dividing polygon; we shall call this an incomplete bounded polyhedron. The other part with no bounding edge will be called an incomplete marginless polyhedron, or a polyhedron with one null rim. For each of these the value of $N_2 - N_1 + N_0$ is 1, the value of $N_1 - N_0$ for the dividing polygon being zero. Similarly when a simply-connected polytope $(Po)_n$ is severed along a simply-connected polytope $(Po)_{n-1}$ (not in general lying in an S_{n-1}), the two parts are incomplete polytopes, one bounded, the other marginless.

10. Euler's Theorem for Simply-connected Polytopes. We shall now prove by induction that *for an Eulerian polytope* $(Po)_n$

$$\phi_n = N_{n-1} - N_{n-2} + \ldots + (-)^{n-1} N_0 + (-1)^n = 1,$$

and for an incomplete bounded polytope with one free rim ϕ_n *has the value* 0.

Let us assume that these results are true up to $n - 1$. Then for a single polytope $(Po)_{n-1}$ we have $N_{n-1} = 1$ while $N_{n-2} - \ldots + (-)^{n-2}N_0 + (-1)^{n-1} = 1$, and therefore $\phi_n = 0$. Let polytopes $(Po)_{n-1}$ be built on, one at a time, so as to produce at any stage an incomplete bounded polytope $(Po)_n$ whose margin is a simply-connected skew polytope $(Po)_{n-1}$. When one more polytope $(Po)_{n-1}$ is applied, we have then added an incomplete marginless polytope $(Po)_{n-1}$; the value of $N'_{n-3} - N'_{n-4} + \ldots + (-)^{n-3}N'_0$ for the margin is by our assumption equal to $1 - (-1)^{n-2}$, hence the value of $N'_{n-2} - N'_{n-3} + \ldots + (-)^{n-2}N'_0$ for the marginless polytope added is $(-1)^n + 1 - (-1)^{n-2} = 1$. We have therefore added 1 to N_{n-1} and also 1 to $N_{n-2} - N_{n-3} + \ldots$, and therefore the value of ϕ_n remains unaltered and equal to 1. The theorem is therefore proved. We have proved also that *for an incomplete marginless polytope with one null rim*

$$N_{n-1} - N_{n-2} + \ldots + (-)^{n-1} N_0 = 1.$$

Ex. Prove that for an incomplete polytope with p free rims, $\phi_n = 1 - p$; and for an incomplete polytope with q null rims, $\phi_n = 1 + (-)^n q$.

11. Other Relations Connecting the number of Boundaries. Euler's equation connects the *total* configurational numbers of a polytope, without reference to the nature of the boundaries, triangular, quadrilateral, or otherwise. In addition there are relations connecting the numbers of boundaries of different forms.

Let us confine our attention first to three dimensions. Let F_n be the number of faces having n sides. The maximum number of sides of a face is V − 1, which occurs in the case of a pyramid only, and the minimum is 3. Hence

$$F_3 + F_4 + \dots + F_{V-1} = F. \quad . \quad . \quad (11\cdot1)$$

Let V_p be the number of vertices at which p edges meet. The maximum value of p is F − 1, which occurs also only in the case of a pyramid, and the minimum value is 3. Hence

$$V_3 + V_4 + \dots + V_{F-1} = V. \quad . \quad . \quad (11\cdot2)$$

Also, since each edge adjoins two faces, we have

$$3F_3 + 4F_4 + \dots + (V-1)F_{V-1} = 2E; \quad . \quad (11\cdot3)$$

and since each edge joins two vertices,

$$3V_3 + 4V_4 + \dots + (F-1)V_{F-1} = 2E. \quad . \quad (11\cdot4)$$

We have also Euler's equation

$$F - E + V = 2. \quad . \quad . \quad . \quad (11\cdot5)$$

Multiplying (11·1) and (11·2) each by 4 and subtracting the sum of (11·3) and (11·4) we get

$$(F_3+V_3) - (F_5+V_5) - 2(F_6+V_6) - \dots = 4(F+V-E) = 8.$$

Hence $$F_3 + V_3 \gtrless 8.$$

The tetrahedron, octahedron, and cube are examples of polyhedra for which $F_3 + V_3 = 8$.

Multiply (11·1) by 3 and subtract from (11·3), and (11·2) by 3 and subtract from (11·4), and we get

$$\rho_F = F_4 + 2F_5 + 3F_6 + \dots + (V-4)F_{V-1} = 2E - 3F,$$
$$\rho_V = V_4 + 2V_5 + 3V_6 + \dots + (F-4)V_{F-1} = 2E - 3V,$$

whence

$$\rho_F + \rho_V = 4E - 3(F+V) = E - 6.$$

For a triangular polyhedron $\rho_F = 0$ and $\rho_V = E - 6$; for a trihedral polyhedron $\rho_V = 0$ and $\rho_F = E - 6$.

For a trihedral polyhedron we have

$$3V = 2E \text{ and } 2F = 4 + 2E - 2V = 4 + V;$$

also, multiplying (11·1) by 6 and subtracting from (11·3),

$$3F_3 + 2F_4 + F_5 - F_7 - 2F_8 - \dots - (V-7)F_{V-1} = 6F - 2E = 12.$$

Putting F_3, F_4, F_7, F_8, etc., = 0 we deduce that *if a trihedral polyhedron has only pentagonal and hexagonal faces the number of pentagonal faces is* 12.

Also for any polyhedron, subtracting (11·3) from 6 times (11·1),

$$3F_3+2F_4+F_5-F_7-2F_8-\ldots-(V-7)F_{V-1}=4E-6V+12 \gtrless 12,$$

since $2E \gtrless 3V$. Hence *there is no convex polyhedron whose faces are all hexagons or polygons with more than* 6 *sides.*

Ex. 1. If a trihedral polyhedron has only quadrilateral and hexagonal faces show that the number of quadrilateral faces is 6.

Ex. 2. If a trihedral polyhedron has only triangular and hexagonal faces show that the number of triangular faces is 4.

12. In space of four and more dimensions the relations between the partial configurational numbers of polytopes become of great complexity owing to the great variety of polytopes of lower dimensions of which their boundaries are composed. There is also a new condition introduced on passing from three to four dimensions due to the necessity that two adjacent polyhedra should have in common a face with the same number of sides; a tetrahedron cannot be adjacent to a cube, for example. The corresponding condition in three dimensions involves only the equality of edges, which is a metrical condition and therefore irrelevant from the point of view of analysis situs. A distinction between odd and even values of n must be emphasised here also. When the dimensional number n is odd Euler's relation is not homogeneous in the total configurational numbers, but when n is even the equation is homogeneous and gives only a relation between the *ratios* of the numbers. As the configurational relations are also homogeneous it would be impossible, in space of four dimensions, for example, to derive from these relations alone results relating to the actual numbers of boundaries of given type such as we found for polyhedra in three dimensions.

13. We conclude this section with some general theorems relating to simplex-polytopes and simplex-polycoryphas.

In a simplex-polytope the boundaries of all dimensions are simplexes; in particular the two-dimensional boundaries are triangles; the three-dimensional boundaries are tetrahedra. In a simplex-polycorypha the angle-constituents of all dimensions are simplexes, i.e. through every $(n-p)$-dimensional boundary there are p boundaries of $n-1$ dimensions, ${}_pC_r$ boundaries of

$n - r$ dimensions; at every vertex there are n edges, no r of which lie in an $(r - 1)$-flat. Hence an r-boundary cannot have more than r edges at a vertex and must therefore be a simplex-polycorypha. Thus *in a simplex-polycorypha the boundaries of all dimensions are simplex-polycoryphas.* Similarly in a simplex-polytope the angle-constituents of all dimensions are simplex-polytopes.

A simplex-polytope which is also a simplex-polycorypha is necessarily a simplex. But, further, a simplex-polycorypha whose boundaries of $n - 2$ dimensions are simplexes is also necessarily a simplex, for its $(n - 1)$-dimensional boundaries are simplex-polycoryphas whose boundaries are simplexes, and are therefore simplexes; the figure is therefore also a simplex-polytope, and therefore a simplex. By induction we can prove finally that *a simplex-polycorypha whose two-dimensional boundaries are all triangles is necessarily a simplex;* and reciprocally, *a simplex-polytope which has just three $(n - 1)$-dimensional boundaries through each boundary of $n - 3$ dimensions is necessarily a simplex.*

14. Proofs of Euler's Theorem in Four Dimensions by Consideration of Angles. Let us return now to examine whether Steiner's and Legendre's methods of proof of Euler's theorem can be extended to higher dimensions.

In space of four dimensions, projecting a convex polytope orthogonally we obtain a polyhedron filled twice over with a honeycomb of polyhedra. Let N'_2, N'_1, N'_0 be the numbers of faces, edges, and vertices of the outer polyhedron, N_3, N_2, N_1, N_0 the numbers of cells, faces, edges, and vertices of the polytope, and $N_2^{(i)}$ the number of faces of the various cells ($i = 1, 2, \ldots, N_3$). Instead of plane angles we have to consider dihedral angles and polyhedral angles (solid angles at a vertex). A solid angle is measured by the area which it cuts out on the surface of a unit sphere whose centre is at the vertex, and the measure of the total solid angle at a vertex is taken to be 4π. Similarly a dihedral angle may be measured by the area which it cuts out on the surface of a unit sphere whose centre lies on the edge, and we shall take the measure of the total dihedral angle at an edge also 4π. Let now $S_1^{(i)}$ denote the sum of the dihedral angles of a polyhedron, $S_0^{(i)}$ the sum of the polyhedral angles, and let S_1', S_0' denote the same for the outer polyhedron. We have then, summing for all the polyhedra into which the outer polyhedron is divided,

$$\Sigma S_1^{(i)} = 2S_1' + 4\pi(N_1 - N_1'), \quad . \quad . \quad (14\cdot1)$$

$$\Sigma S_0^{(i)} = 2S_0' + 4\pi(N_0 - N_0'). \quad . \quad . \quad (14\cdot2)$$

We have next to get a relation between S_1 and S_0 for any convex polyhedron. Construct a small sphere round any vertex A; let α_0 be the measure of the solid angle, and $\alpha_1, \beta_1, \ldots$ those of the dihedral angles at the edges through A. We have then on the surface of the sphere a spherical polygon of area α_0, and angles $\alpha_1, \beta_1, \ldots$, the measure of the angles being such that the total angle round a point on the surface of the sphere is 4π (not 2π). If n is the number of sides of this spherical polygon we have

$$\alpha_0 = \tfrac{1}{2}\Sigma\alpha_1 - (n - 2)\pi.$$

Summing for all the vertices we have

$$S_0^{(i)} = S_1^{(i)} - \Sigma(n - 2)\pi.$$

But $\Sigma n = 2N_1^{(i)}$, hence

$$S_1^{(i)} - S_0^{(i)} = 2\pi N_1^{(i)} - 2\pi N_1^{(i)} = 2\pi(N_2^{(i)} - 2). \,. \quad (14\cdot3)$$

Subtracting equation (14·2) from (14·1) and applying (14·3) to both accented and non-accented symbols, we have

$$\Sigma S_1^{(i)} - \Sigma S_0^{(i)} = \Sigma(N_2^{(i)} - 2)2\pi = (N_2' - 2)4\pi + 4\pi(N_1 - N_0 - N_1' + N_0').$$

But $\Sigma N_2^{(i)} = 2N_2$, hence, dividing by 4π we have

$$N_2 - N_3 = N_2' - 2 + N_1 - N_0 - N_1' + N_0',$$

therefore

$$N_3 - N_2 + N_1 - N_0 = 2 - N_0' + N_1' - N_2' = 0,$$

by Euler's theorem in three dimensions.

Legendre's proof is still more easily extended. Projecting a convex polytope from an interior point O on to a unit hypersphere with centre O, the hypersphere is divided into a honeycomb of spherical polyhedra. The relation between the angle-sums of a spherical polyhedron is the same as for a euclidean polyhedron, viz.:

$$S_1 - S_0 = 2\pi(n_2 - 2),$$

where n_2 is the number of two-dimensional faces

Summing for all the cells of the honeycomb we have

$$4\pi N_1 - 4\pi N_0 = 2\pi\Sigma(n_2^{(i)} - 2)$$
$$= 4\pi N_2 - 4\pi N_3.$$

hence $$N_3 - N_2 + N_1 - N_0 = 0.$$

This connection between Euler's equation and angle-sums can be extended to n dimensions. Euler's equation for $(\text{Po})_n$ is definitely connected with a relation connecting the angle-sums of a $(\text{Po})_{n-1}$ in spherical space.

15. Measurement of n-Dimensional Angles. We have first to explain the different species of angles and their measure, and it is necessary to consider non-euclidean as well as euclidean geometry. In a plane there is just one species of angle, the plane angle. With the radian as unit, the complete angle at a point is 2π, and the measure of a right angle is $\frac{1}{2}\pi$. This is true both in euclidean and in non-euclidean geometry, although in the latter case the unit angle cannot be defined as the angle subtended by an arc equal to the radius, for in non-euclidean geometry the ratio of the circumference to the radius is not constant. However, for any given angle it is true, both in euclidean and in non-euclidean geometry, that the ratio of the arc to the whole circumference, or the ratio of the sector to the area of the whole circle, is independent of the radius. This ratio therefore affords a measure for a plane angle in non-euclidean as well as in euclidean geometry, and 2π times this measure is the "radian" measure.

In three dimensions there are, in addition to the plane angle, two species of "solid" angles: dihedral angles, or angles at an edge; and polyhedral angles, or angles at a vertex. To measure a solid angle, let O be the vertex, or, in the case of a dihedral angle, any point on the edge; and construct a sphere with centre O. The plane boundaries of the angle cut out a certain area on the surface of the sphere, and, both in euclidean and in non-euclidean geometry, the ratio of this area to the whole surface of the sphere is independent of the length of the radius. The unit of solid angle is chosen so that, in the case of euclidean geometry, the measure of the angle is equal to the ratio of the area cut out on the surface of the sphere to the square of the radius; the measure of the complete solid angle, at a vertex or at an edge, is then 4π. We may therefore define the "radian" measure of a solid angle, in euclidean or non-euclidean geometry, as 4π times the ratio of the area cut out on the surface of the sphere to the whole area of the sphere, or 4π times the ratio of the volume of the sector to the whole volume of the sphere.

In spherical or elliptic geometry, where the whole of space is finite, there is another way of defining the radian measure. Consider first spherical geometry of two dimensions which is the same as the geometry on a sphere, great circles taking the place of straight lines. The angle at a point, the angle between two great circles, can be measured by the ratio of the area (lune) enclosed between the two great semi-circles to the whole surface of the sphere. If k is the radius of the sphere, or the "space-constant" of the spherical space, the whole area of the two-dimensional space is $4\pi k^2$. In radian measure the whole angle at a point is 2π, hence in spherical geometry of two dimensions the angle between two rays is equal to the ratio of the area enclosed between the two rays to $2k^2$.

Elliptic geometry differs from spherical geometry in the circumstance that antipodal points always coincide. The whole area of the elliptic "plane" is $2\pi k^2$. Two rays which proceed from a point O do not meet until they return to O. The angular region enclosed between two opposite rays comprises the whole of the elliptic plane, and the radian measure of this angle is π. Hence, as in spherical geometry, the radian measure of a plane angle in elliptic geometry is equal to the ratio of the area enclosed between the two rays to $2k^2$.

In n dimensions there are $n - 1$ species of angles: angles at a point, a line, a plane, . . ., an $(n - 2)$-flat, or, enumerating them in the reverse order, angles bounded by 2, . . ., $n - 2$, $n - 1$, or more than $n - 1$ hyperplanes. In euclidean geometry we define the radian measure of an n-dimensional angle as the ratio of the surface-content cut out of a hypersphere whose centre is on the axis (vertex, edge, etc.) to the $(n - 1)$th power of the radius. The surface-content of a hypersphere of radius k in n dimensions is $k^{n-1}2\pi^{\frac{1}{2}n}/\Gamma(\frac{1}{2}n)$. Hence the radian measure of the complete angle at a point, or an edge, etc., is $2\pi^{\frac{1}{2}n}/\Gamma(\frac{1}{2}n)$. Hence, in both euclidean and non-euclidean geometry, the radian measure of an n-dimensional angle is $2\pi^{\frac{1}{2}n}/\Gamma(\frac{1}{2}n)$ times the ratio of the surface-content cut out of a hypersphere whose centre is on the axis to the surface-content of the whole hypersphere; or the corresponding ratio of volume-contents. In spherical geometry, taking the hypersphere in the extreme case when it occupies the whole of space, whose volume-content, which is the surface-content of a hypersphere of radius k of $n + 1$ dimensions, is $2\pi^{\frac{1}{2}(n+1)}k^n/\Gamma\{\frac{1}{2}(n + 1)\}$, the radian measure of an angle is $2\pi^{\frac{1}{2}n}/\Gamma(\frac{1}{2}n)$ times the ratio of the volume-content enclosed by the bounding hyperplanes to $2\pi^{\frac{1}{2}(n+1)}k^n/\Gamma\{\frac{1}{2}(n + 1)\}$, i.e. equal to the ratio of this volume-content to $k^n\sqrt{\pi} . \Gamma\{\frac{1}{2}(n+1)\}/\Gamma(\frac{1}{2}n)$. This holds also for elliptic geometry. The radian measure of the complete angle is $2\pi^{\frac{1}{2}n}/\Gamma(\frac{1}{2}n)$, and the total volume of spherical space is $2\pi^{\frac{1}{2}(n+1)}k^n/\Gamma\{\frac{1}{2}(n + 1)\}$.

16. Relation Between the Area and Angle-Sum of a Spherical Triangle. We proceed now to investigate the relations connecting the angle-sums of a simplex in spherical geometry with space-constant k.

For two dimensions we have the familiar relation connecting the sum of the angles and the area of a spherical triangle. Let V be the area, and α, β, γ the radian measures of the angles, the complete angle at a point being 2π. We have (Fig. 41)

$$\alpha = 2\pi \frac{\text{area of lune ABA'C}}{\text{whole area of spherical surface}}.$$

Now the six lunes ABA'C, AB'A'C', etc., cover the whole sphere once with the exception of the interior of the triangle ABC and the antipodal triangle A'B'C' which are covered three times.

Hence the sum of the areas of the six lunes

$$= 4\pi k^2 + 4V$$
$$= 2k^2 . 2\Sigma\alpha.$$

Hence

$$V = k^2 (\Sigma\alpha - \pi).$$

17. Angular Regions of a Simplex. The same method can be applied to a simplex in spherical space of n dimensions. The $n + 1$ hyperplanes which form the faces of the simplex divide the whole spherical space into 2^{n+1} regions, which occur in pairs of antipodal regions. Each of these regions is the interior of a simplex. Denote the vertices of the chosen simplex by 0, 1, . . ., n, and those of the antipodal simplex by $0'$, $1'$, . . ., n'. Then the vertices of each of the 2^{n+1} simplexes are represented by the $n+1$ digits 0, 1, . . ., n, with or without accents. It is convenient to represent the interior of

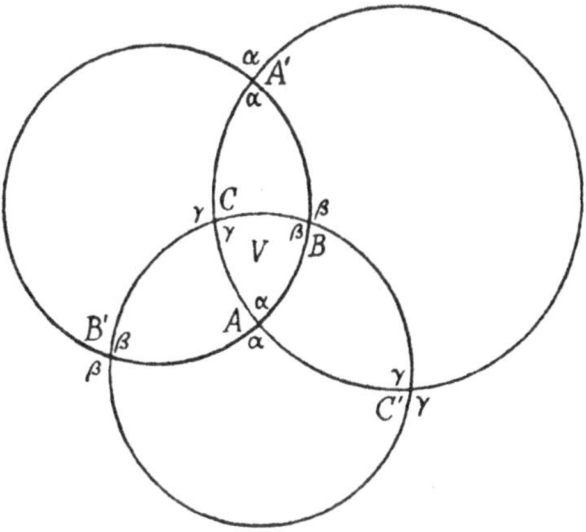

FIG. 41.

each region by the unaccented figures alone. Thus the antipodal regions ($0'12 \ldots n$) and ($01'2' \ldots n'$) will be represented by ($12 \ldots n$) and (0) respectively; this indicates their relationship to the given simplex as standing on an $(n-1)$-dimensional boundary and a vertex respectively. ($01 \ldots n$) is the interior of the chosen simplex, and ( ) that of the antipodal simplex.

The number of regions of the type ($01 \ldots r$), standing on an r-dimensional boundary, is ${}_{n+1}C_{r+1}$, and there are the same number of regions of the type ($r+1, \ldots, n$). When n is even, antipodal regions are always of different type, but when n is odd the antipodal regions $\{01 \ldots \frac{1}{2}(n-1)\}$ and $\{\frac{1}{2}(n+1) \ldots n\}$ are of the same type, both standing on boundaries of $\frac{1}{2}(n-1)$ dimensions. Any region of type ($01 \ldots r$) will be denoted by $[r+1]$.

18. We have now to consider the angular regions. A single hyperplane $123 \ldots n$ divides space into two regions; that which contains the interior of the simplex will be taken as the positive side or interior and denoted by $\alpha_{12\ldots n}$, the other region, which is antipodal to this, will be called the co-interior and denoted by $\alpha'_{12\ldots n}$. Two hyperplanes $123\ldots n$ and $023\ldots n$ determine an angular region whose interior is that which contains the interior of the simplex and will be denoted by $\alpha_{23\ldots n}$; the co-interior, which contains the antipodal simplex, will be denoted by $\alpha'_{23\ldots n}$; and so on. Thus $\alpha_{pqr\ldots}$ represents an angular region whose axis is the boundary of the simplex determined by the points $p, q, r, \ldots$ $\alpha_{012\ldots n}$ can be taken to mean the whole of space $= S$; $\alpha_{12\ldots n} = \frac{1}{2}S$. α will be taken to mean the volume V of the simplex, α' that of the antipodal simplex.

The angular region $\alpha_{r+1,\ldots,n}$ contains all the regions $[k]$ whose symbols include all the digits $0, 1, \ldots, r$ which $\alpha_{r+1,\ldots,n}$ does not contain; i.e. it contains the following regions:—

- the simplex $(01\ldots n)$,
- ${}_{n-r}C_1$ regions on $(n-1)$-dimensional boundaries, whose symbols are formed with n of the digits $0, 1, \ldots, n$ always including $0, 1, \ldots, r$;
- ${}_{n-r}C_2$ regions on $(n-2)$-dimensional boundaries, and so on, up to the region $(01\ldots r)$ on the r-dimensional boundary.

The number of angular regions of the type $\alpha_{r+1,\ldots,n}$ is ${}_{n+1}C_{r+1} = {}_{n+1}C_{n-r}$.

We have next to consider the sums of the angular regions of the same type. Let $\Sigma[r]$ be denoted by A_r, so that A_{n+1} denotes the volume of the simplex V, and A_0 the (equal) volume of the antipodal simplex. Let A'_r denote the sum of the regions antipodal to those which compose A_r. Then $A'_{r+1} = A_{n-r}$, both in total content and in separate parts, and $A_{r+1} = A_{n-r}$ in content.

Let $\Sigma\alpha_{01\ldots r}$, the summation extending to all angular regions with $r+1$ suffixes, be denoted by Σ_r, and $\Sigma\alpha'_{01\ldots r}$ by Σ'_r. $\Sigma\alpha(=\alpha)$ may be denoted by Σ_{-1}, so that $\Sigma_{-1} = V$. $\Sigma\alpha_{01\ldots n} = \alpha_{01\ldots n} = \Sigma_n =$ the whole content of space S. $\Sigma\alpha_{12\ldots n} = \Sigma_{n-1} = \frac{1}{2}(n+1)S = \frac{1}{2}(n+1)\Sigma_n$.

19. Now a given region $[s]$ $(s \geqq n-r)$ is contained in each of the angular regions of type $\alpha_{01\ldots r}$ whose suffixes contain all the $n+1-s$ digits which are not included in the symbol of $[s]$, and the number of these angular regions is ${}_sC_{n-r}$; hence in the sum $\Sigma\alpha_{01\ldots r}$ each region $[s]$ occurs ${}_sC_{n-r}$ times. Hence

$$\begin{aligned}\Sigma_r &= {}_{n+1}C_{n-r}A_{n+1} + {}_nC_{n-r}A_n + \ldots + {}_{n-r}C_{n-r}A_{n-r} \\ &= {}_{n+1}C_{r+1}V + {}_nC_rA_1 + \ldots + {}_{n-r}C_0A_{r+1}.\end{aligned}$$

Thus, putting $r = 0, 1, \ldots, n$, we have the following $n+1$ equations in $A_1, \ldots, A_n$:

$$\begin{aligned}\Sigma_0 &= {}_{n+1}C_1V + A_1,\\ \Sigma_1 &= {}_{n+1}C_2V + {}_nC_1A_1 + A_2,\\ &\ldots\ldots\ldots\ldots\\ \Sigma_{n-1} &= {}_{n+1}C_nV + {}_nC_{n-1}A_1 + {}_{n-1}C_{n-2}A_2 + \ldots + A_n,\\ \Sigma_n &= V + A_1 + A_2 + \ldots + A_n + V.\end{aligned}$$

Also the relations

$$A_{r+1} = A_{n-r} \qquad (r=0, 1, \ldots, \tfrac{1}{2}n-1 \text{ or } \tfrac{1}{2}(n-1))$$

supply $\frac{1}{2}n$ or $\frac{1}{2}(n+1)$ further equations, according as n is even or odd. Hence by eliminating the A's we get, connecting the volume V and the sums of the angular regions Σ_r, $\frac{1}{2}n+1$ or $\frac{1}{2}(n+1)$ relations, i.e. $[\frac{1}{2}n]+1$, where $[\frac{1}{2}n]$ denotes the integral part of $\frac{1}{2}n$.

20. Relation between the Volume and Angle-sums of a Polytope. One of these relations is

$$\Sigma_n - \Sigma_{n-1} + \ldots + (-)^n\Sigma_0 = V + (-)^nV,$$

as is easily verified since the coefficient of A_{n-r} is $(1-1)^{r+1} = 0$. The other relations will not be required.

We shall now prove that this relation holds not only for simplexes, but for any polytope which can be divided into simplexes. Consider first a simplex-polytope and divide it *centrally;* i.e. take any point in its interior and join it to all the vertices. We assume that the joining rays lie entirely within the polytope, and the polytope is thus divided into simplexes. Let Σ_r denote the sum of the angular regions at the r-boundaries for the polytope and Σ'_r the corresponding sum for a constituent simplex; V the whole volume and V′ that of a constituent simplex. Then

$$\begin{aligned}
\Sigma\Sigma'_n &= N_{n-1}\Sigma_n, \\
\Sigma\Sigma'_{n-1} &= \Sigma_{n-1} + N_{n-2}\Sigma_n, \\
\Sigma\Sigma'_{n-2} &= \Sigma_{n-2} + N_{n-3}\Sigma_n, \\
&\cdot \quad \cdot \quad \cdot \quad \cdot \quad \cdot \\
\Sigma\Sigma'_1 &= \Sigma_1 + N_0\Sigma_n, \\
\Sigma\Sigma'_0 &= \Sigma_0 + \Sigma_n, \\
\Sigma V' &= V.
\end{aligned}$$

For each constituent simplex

$$\Sigma'_n - \Sigma'_{n-1} + \ldots + (-)^n\Sigma'_0 = V' + (-)^n V'.$$

Hence

$$\begin{aligned}
\Sigma\{V' + (-)^n V'\} = \{N_{n-1} - N_{n-2} + \ldots + (-)^{n-1}N_0 + (-1)^n\}\Sigma_n \\
- \{\Sigma_{n-1} - \Sigma_{n-2} + \ldots - (-)^n\Sigma_0\}.
\end{aligned}$$

But by Euler's theorem the coefficient of $\Sigma_n = 1$, hence

$$\Sigma_n - \Sigma_{n-1} + \ldots + (-)^n\Sigma_0 = V + (-)^n V.$$

Next consider any Eulerian polytope; take any r-boundary and divide it centrally. Let α_r be the volume of the angular region at that r-boundary as edge; Σ_s the sum of the angular regions at an s-boundary for the given polytope and Σ'_s that for the transformed polytope. Then

$$\begin{aligned}
\Sigma'_s &= \Sigma_s, \quad (s = n, n-1, \ldots, r+1) \\
\Sigma'_r &= \Sigma_r + (N_{r-1,r} - 1)\alpha_r, \\
\Sigma'_s &= \Sigma_s + N_{s-1,r}\alpha_r, \quad (s = r-1, r-2, \ldots, 1) \\
\Sigma'_0 &= \Sigma_0 + \alpha_r, \\
V' &= V.
\end{aligned}$$

Hence $\qquad \Sigma'_n - \Sigma'_{n-1} + \ldots + (-)^n\Sigma'_0$

$$\begin{aligned}
&= \Sigma_n - \Sigma_{n-1} + \ldots + (-)^{n-r}\Sigma_r + \ldots + (-)^n\Sigma_0 \\
&+ (-)^{n-r}\{N_{r-1,r} - N_{r-2,r} + \ldots + (-)^{r-1}N_{0,r} + (-1)^r - 1\}\alpha_r.
\end{aligned}$$

But the given r-boundary being assumed to be an Eulerian polytope,

$$N_{r-1,r} - N_{r-2,r} + \ldots + (-)^{r-1}N_{0,r} + (-1)^r = 1.$$

Hence the expression $\Sigma_n - \Sigma_{n-1} + \ldots + (-)^n\Sigma_0$ is not altered by this transformation. If now all the boundaries of the polytope, of all dimensions, are divided centrally, the polytope is transformed into a simplex-polytope. Hence the relation

$$\Sigma_n - \Sigma_{n-1} + \dots + (-)^n\Sigma_0 = \{1 + (-1)^n\}V \quad (20{\cdot}1)$$

is true for any Eulerian polytope in spherical hyperspace.

21. Now let S_r denote the sum of all the angles at r-dimensional edges expressed in radian measure, so that

$$\Sigma_r = S_r \,.\, k^n \,.\, \frac{\Gamma(\frac{1}{2}n)\pi^{\frac{1}{2}}}{\Gamma\{\frac{1}{2}(n+1)\}}, \quad (r = 0, 1, \dots, n-2)$$

$$\Sigma_n = k^n \frac{2\pi^{\frac{1}{2}(n+1)}}{\Gamma\{\frac{1}{2}(n+1)\}},$$

$$2\Sigma_{n-1} = N_{n-1}\,\Sigma_n.$$

Then equation (20·1) becomes

$$k^n\left\{S_{n-2} - S_{n-3} + \dots + (-)^n S_0 + \frac{\pi^{\frac{1}{2}n}}{\Gamma(\frac{1}{2}n)}(2 - N_{n-1})\right\}$$
$$= \frac{\Gamma\{\frac{1}{2}(n+1)\}}{\Gamma(\frac{1}{2}n)\sqrt{\pi}}\{1 + (-1)^n\}V. \quad (21{\cdot}1)$$

This relation is true for any Eulerian polytope in spherical or elliptic space of n dimensions. In euclidean space $k \to \infty$ and the equation reduces to

$$S_{n-2} - S_{n-3} + \dots + (-)^n S_0 = \frac{\pi^{\frac{1}{2}n}}{\Gamma(\frac{1}{2}n)}(N_{n-1} - 2). \quad (21{\cdot}2)$$

In hyperbolic space k is purely imaginary. Replacing k by ik when n is even we have

$$-k^n\left\{S_{n-2} - S_{n-3} + \dots + (-)^n S_0 + \frac{\pi^{\frac{1}{2}n}}{\Gamma(\frac{1}{2}n)}(2 - N_{n-1})\right\}$$
$$= \frac{\Gamma\{\frac{1}{2}(n+1)\}}{\Gamma(\frac{1}{2}n)\sqrt{\pi}}\{1 + (-1)^n\}V \quad . \quad (21{\cdot}3)$$

When n is odd the right-hand side vanishes; hence in this case equation (21·2) is true in both euclidean and non-euclidean geometry.

Ex. 1. If α_0 is the solid angle at a vertex and α_1 the dihedral angle at an edge, show that for the regular tetrahedron, cube, and dodecahedron $3\alpha_1 - 2\alpha_0 = 2\pi$, for the octahedron $2\alpha_1 - \alpha_0 = 2\pi$, and for the icosahedron $5\alpha_1 - 2\alpha_0 = 6\pi$.

Ex. 2. For the regular polytopes in S_4 prove that

$$N_{31}(\alpha_2 - \pi^2) = 2(\alpha_1 - \pi^2).$$

REFERENCES

CAUCHY, A. L. Recherches sur les polyèdres. J. Éc. Polyt., **9**, cah. xvi. (1813), pp. 68-86-98 ; Oeuvres (2), i.

DEHN, M. Die Eulersche Formel im Zusammenhang mit dem Inhalt in der nichteuklidischen Geometrie. Math. Ann., **61** (1905), 561-586.

— M., and HEEGAARD, P. Analysis situs. Encykl. math. Wiss., III., 1_1 (III., AB 2), 1907.

EULER, L. Elementa doctrinae solidorum.—Demonstratio nonnullarum insignium proprietatum, quibus solida hedris planis inclusa sunt praedita. Novi comment acad. sc. imp. Petropol., **4** (1752-3), 109-140-160.

LEGENDRE, A. M. Éléments de géométrie. Paris, 1794. (Liv. vii., Prop. xxv. and Note viii.)

LHUILIER, S. A. J. Mémoire sur la polyédrométrie, contenant une démonstration directe du théorème d'Euler sur les polyèdres, et un examen de diverses exceptions auxquelles ce théorème est assujetti. Gergonne Ann. Math., **3** (1812), p. 169.

LISTING, J. B. Der Census räumlicher Complexe oder Verallgemeinerung des Euler'schen Satzes von den Polyedern. Göttingen, 1862. (Gött. Ab. Ges. Wiss., **10**.)

POINCARÉ, H. Analysis situs. J. Éc. Polyt. (2) **1** (1895). Continued in Palermo Rend., **13** (1899); Proc. London Math. Soc., **37** (1900); Bull. Soc. Math., **30** (1902); Liouville J. Math. (5) **8** (1902); and Palermo Rend., **18** (1904).

SOMMERVILLE, D. M. Y. The relations connecting the angle-sums and volume of a polytope in space of n dimensions. Proc. R. Soc., London, A, **115** (1927), 103-119.

v. STAUDT, G. K. C. Geometrie der Lage. Nürnberg, 1847. (§ 4, p. 20.)

STEINER, J. Leichter Beweis eines stereometrischen Satzes von Euler. Crelle J. Math., **1** (1826), 364-367. Werke, i., p. 97.

CHAPTER X

THE REGULAR POLYTOPES

1. THE regular polyhedra have been famous from the time of the Greek geometers, and it will be well, before considering the corresponding figures in higher space, to examine these more familiar figures in some detail. Each is characterised by two conditions: (1) its faces are regular polygons with the same number of sides, the lengths of the sides being therefore all equal; and (2) there are the same number of faces and edges at each vertex. It can be proved that the vertices are then also necessarily regular, i.e. not only the face-angles but also the dihedral angles are all equal. Notice that condition (1) by itself is not sufficient; e.g. two equal regular tetrahedra placed base to base form a polyhedron whose faces are all equilateral triangles, but it is nevertheless not a regular polyhedron. From the point of view of morphology, or analysis situs, the regular polyhedra are special forms of a more general class, the *homogeneous* polyhedra, with which they are isomorphic. Thus the regular tetrahedron is isomorphic with the general tetrahedron, the fact that the faces are all equilateral triangles being an unnecessary accident. To determine the homogeneous polyhedra we leave out the condition that the faces are regular polygons and define them by the two conditions: (1) each face has the same number, n say, of edges or vertices, and (2) each vertex has the same number, p say, of edges or faces.

2. **Polyhedral Configurations.** A homogeneous polyhedron is therefore a particular type of configuration which is characterised by a matrix of the type

$$\begin{array}{|ccc|} N_0 & 2 & n \\ p & N_1 & n \\ p & 2 & N_2 \end{array}$$

The two numbers $N_{02} = N_{12} = n$ (the number of sides or vertices of the polygons), and $N_{10} = N_{20} = p$ (the number of edges or faces at a vertex), are called its *fundamental numbers.*

We have then the relations

$$pN_0 = 2N_1 = nN_2,$$

and Euler's formula

$$N_2 - N_1 + N_0 = 2;$$

whence $$N_0 = 2n\lambda,\ N_1 = np\lambda,\ N_2 = 2p\lambda$$

where $$\lambda = \frac{2}{2(n + p) - np}.$$

3. Elliptic, Euclidean, and Hyperbolic Types. There are now three cases.

(i) If the polyhedron is finite, λ must be finite and positive, hence

$$p < \frac{2n}{n - 2}.$$

The smallest value of n is 3, and then $p < 6$; for $n = 4, p < 4$; for $n = 5, p < 4$; for $n \geq 6, p < 3$, but 3 is the least value of p. Hence we have just the five cases

n	3	3	3	4	5
p	3	4	5	3	3

(ii) If $2(n + p) - np = 0$, λ becomes infinite, and we have an "open" polyhedron. The only cases here are

n	3	4	6
p	6	4	3

(iii) If $2(n + p) - np < 0$, no configuration appears at first sight to be possible. We shall examine this presently.

In the first case the polyhedron may be represented isomorphically by a configuration of N_0 points on the surface of a sphere, and if the edges are replaced by arcs of great circles we have a network on the sphere having N_2 meshes, N_0 nodes, and

N_1 edges. If S denotes the sum of the angles of a spherical n-gon, its area $= \{S - (n - 2)\pi\}r^2$, where r is the radius of the sphere.

Now $$\Sigma S = 2\pi N_0,$$
and $$pN_0 = 2N_1 = nN_2.$$

Hence the total area of the meshes is

$$\{2N_0 - (n - 2)N_2\}\pi r^2 = \frac{2(n + p) - np}{n} N_0 \pi r^2.$$

If $2(n + p) - np = 0$, either r, or N_0 (and therefore N_1 and N_2), or both r and N_0, must become infinite. If r is infinite the sphere becomes a plane. If r is finite the meshes become indefinitely small, and euclidean geometry applies to them since now $S = (n - 2)\pi$. In either case then the configuration can be represented by a network on the euclidean plane.

If $2(n + p) - np < 0$ we would have to postulate a sphere of imaginary radius, since r^2 must be negative. Now the formulæ of spherical trigonometry, when the radius becomes purely imaginary, are still real; e.g. the radius being ik, the formulæ

$$\left.\begin{array}{l}\cos A = \tan \frac{b}{ik} \cot \frac{c}{ik} \\ \cos A = \sin B \cos \frac{a}{ik}\end{array}\right\} \text{ are equivalent to } \left.\begin{array}{l}\cos A = \tanh \frac{b}{k} \coth \frac{c}{k} \\ \cos A = \sin B \cosh \frac{a}{k}\end{array}\right\},$$

and similarly for the others, and the latter are the trigonometrical formulæ which hold in the non-euclidean geometry of Lobachevsky, or hyperbolic geometry; the former, for a real radius (replacing ik by k) being the trigonometrical formulæ for spherical or elliptic geometry.

Case (iii) can then be represented by a network on the hyperbolic plane, case (ii) by a network on the euclidean plane, and case (i) by a network on the elliptic plane.*

* In hyperbolic geometry there are three types of spheres: spheres of real radius, like ordinary spheres; spheres of infinite radius, which are not planes but the so-called "limit-surfaces," which correspond conformally to euclidean planes; and spheres of unreal radius, surfaces equidistant from a plane, "equidistant-surfaces," which include the plane as a particular case. Thus in hyperbolic geometry all three types of networks can be represented.

4. The following table is a summary of the homogeneous plane networks, where k_1 and k_2 are written instead of n and p.

		k_1.	k_2.	N_0.	N_1.	N_2.
Elliptic	Tetrahedral	3	3	4	6	4
	Octahedral	3	4	6	12	8
	Hexahedral (cubical)	4	3	8	12	6
	Icosahedral	3	5	12	30	20
	Dodecahedral	5	3	20	30	12
Euclidean	Triangular	3	6	1 :	3 :	2
	Hexagonal	6	3	2 :	3 :	1
	Square	4	4	1 :	2 :	1
Hyperbolic		3	>6	6 :	$3k_2$:	$2k_2$
		4	>4	4 :	$2k_2$:	k_2
		5	>3	10 :	$5k_2$:	$2k_2$
		6	>3	6 :	$3k_2$:	k_2
		>6	≥ 3	$2k_1$:	k_1k_2 :	$2k_2$

5. The five elliptic networks correspond to the five regular polyhedra. The octahedral and hexahedral networks, the

Tetrahedron.

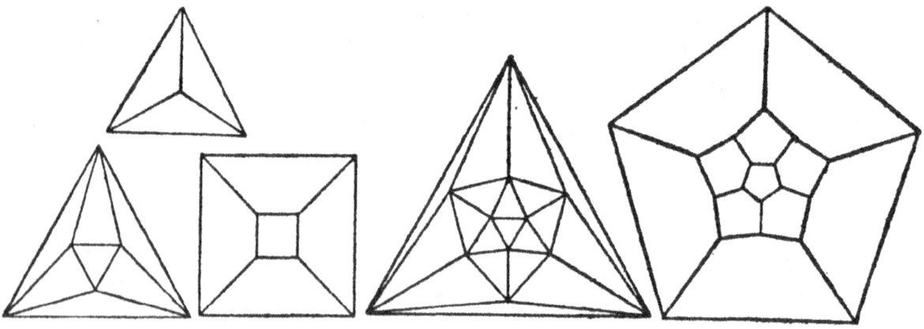

Octahedron. Hexahedron. Icosahedron. Dodecahedron.

FIG. 42.—Schlegel diagrams of the regular polyhedra.

icosahedral and dodecahedral, are reciprocals, the values of k_1 and k_2, N_0 and N_2, being interchanged, N_1 having the same value for both. The tetrahedral network is self-reciprocal. So also the triangular and hexagonal networks are reciprocal, and the square network is self-reciprocal.

The spherical (elliptic) networks are most conveniently represented by the Schlegel diagrams which represent the corresponding polyhedra.

The following diagrams are the exact stereographic projections of the spherical networks, the centre of projection being at the centre of a mesh. All great circles are represented by

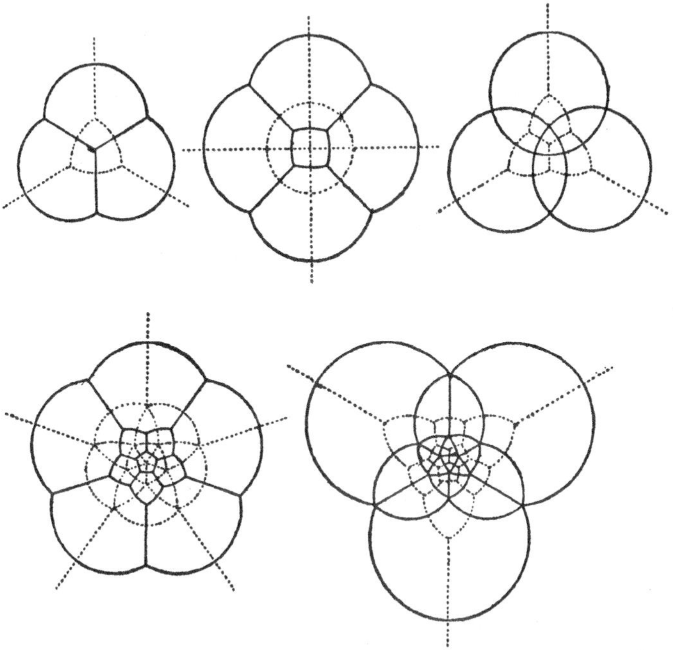

FIG. 43.—Stereographic projections of the regular spherical networks.

circles. On each diagram is represented also the reciprocal network with dotted lines.

6. Homogeneous Polytopes in n Dimensions. We have next to consider the corresponding problem in space of higher dimensions. A homogeneous polytope of n dimensions, like a homogeneous polyhedron in three dimensions, is characterised by two conditions. The first condition is that its $(n - 1)$-boundaries or *face-constituents are homogeneous polytopes, of* $n - 1$

dimensions, all of the same form. The second condition corresponds to the fact that a homogeneous polyhedron has the same number of edges at each vertex, but the condition is not quite so simple. If the elements which pass through a given vertex O are cut by a hyperplane not passing through the vertex, the section is a polytope of $n-1$ dimensions in which any boundary of $r-1$ dimensions is the section of an r-boundary through O of the given polytope. This is a vertex- or angle-constituent. The second condition is then that all the *angle-constituents are homogeneous polytopes of the same form.*

To form a homogeneous polytope of n dimensions we have therefore to choose for the face- and angle-constituents two homogeneous polytopes of $n-1$ dimensions. But we shall see that these two polytopes cannot be chosen independently.

7. The Eleven Homogeneous Honeycombs in Three Dimensions. Let us confine ourselves for the present to four dimensions. We have at our choice five homogeneous polyhedra of three dimensions: tetrahedron, hexahedron, dodecahedron, with 3 edges at a vertex; octahedron with 4; and icosahedron with 5. If the face-constituent is a tetrahedron, hexahedron, or dodecahedron, the angle-constituent must have triangular faces, and must therefore be either a tetrahedron, an octahedron, or an icosahedron. If the face-constituent is an octahedron, the angle-constituent must have quadrilateral faces, and can only be a hexahedron. If the face-constituent is an icosahedron, the angle-constituent must have pentagons for faces, and can only be a dodecahedron. There are thus not more than eleven possible cases. These correspond to homogeneous *honeycombs* in space of three dimensions. We shall express these conditions in a more convenient form.

8. Configurational Numbers. A homogeneous polytope of four dimensions is a configuration, and we shall employ the usual notation for its configurational numbers: N_{pq} = the number of p-boundaries incident with a given q-boundary, N_{pp} or N_p = the total number of p-boundaries, ${}_rN_{pq}$ = the number of p-boundaries passing through a q-boundary and lying in an r-boundary. Denote the configurational numbers of the face-constituents by F_{pq} and those of the vertex-constituents by V_{pq}.

Then $N_{pq} = F_{pq} \quad (p < q < 3)$,
and $N_{p3} = F_p \quad (p = 0, 1, 2)$.
Also $N_{pq} = V_{p-1, q-1} \quad (p > q > 0)$,
and $N_{p0} = V_{p-1} \quad (p = 1, 2, 3)$.

Now in each 3-dimensional boundary the number of lines or planes through a point is F_{10} or F_{20}; and (what is the same thing) of the elements through each vertex the number of lines or planes in each 3-dimensional boundary is V_{02} or V_{12}. Hence $F_{10} = F_{20} = V_{02} = V_{12}$; and if the fundamental numbers of the face-constituents are $F_{02} = F_{12} = N_{02} = N_{12} = k_1$ and $F_{10} = F_{20} = {}_3N_{10} = {}_3N_{20} = k_2$, those of the angle-constituents are $V_{02} = V_{12} = k_2$ and $V_{10} = V_{20} = N_{21} = N_{31} = k_3$. The three numbers k_1, k_2, k_3 may be called the *fundamental numbers* of the four-dimensional configuration.

k_1 = the number of vertices or edges of a polygon,
k_2 = the number of edges or planes through each vertex of a polyhedron,
k_3 = the number of planes or 3-dimensional boundaries through each edge.

9. We shall consider only polytopes whose faces and angles are all finite configurations, so that only the 5 elliptic types in three dimensions are available for their construction. The possible values of k_1, k_2, k_3 are then

333, 334, 335; 343, 353; 433, 434, 435; 533, 534, 535.

But, just as in three dimensions, they may be of three types: elliptic, euclidean, or hyperbolic.

This is a metrical distinction, and the discrimination can be most readily effected by considering the special case of regular polytopes. We can then distinguish them by the number of cells or 3-dimensional boundaries which meet at an edge. These are as follows:—

3, 4, or 5 tetrahedra, hexahedra, or dodecahedra,
3 octahedra, or 3 icosahedra.

It is necessary therefore to determine the dihedral angles of the regular polyhedra in euclidean space of three dimensions. Let A_0, B_0, C_0, . . . (Fig. 44) be the vertices of one face of a polyhedron whose fundamental numbers are k_1, k_2. Let C_1 be the mid-point of an edge through C_0, C_2 the centre of a face through C_0C_1, and C_3 the centre of the polyhedron. The angle $C_0C_2C_1 = \pi/k_1$; let $\angle C_1C_0C_2 = \theta_1$, $\angle C_2C_1C_3 = \theta_2$. Draw a unit sphere round C_0; the edges and faces through C_0 cut this in a regular spherical polygon of k_2 sides; θ_1 is half the length of its edge, and θ_2 half the magnitude of its angle, while π/k_2

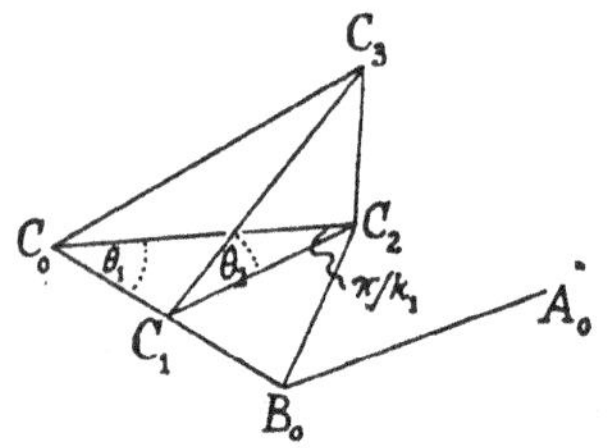

FIG. 44.

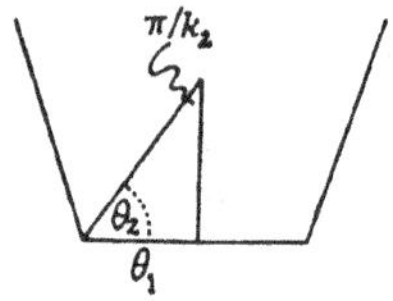

FIG. 45.

is half the angle which an edge subtends at the centre. Hence by spherical trigonometry (Fig. 45)

$$\cos \pi/k_2 = \sin \theta_2 \cos \theta_1 = \sin \theta_2 \sin \pi/k_1,$$

which determines the dihedral angle $2\theta_2$ of the polyhedron.

By calculation we find the following table:

	k_1.	k_2.	$\cos 2\theta_2$.	$2\theta_2$.	Possible Values of k_3.
Tetrahedron .	3	3	$\frac{1}{3}$	70·5°	3, 4, or 5
Octahedron . .	3	4	$-\frac{1}{3}$	109·5	3
Hexahedron . .	4	3	0	90	3, (4)
Icosahedron . .	3	5	$-\frac{\sqrt{5}}{3}$	138·2	—
Dodecahedron .	5	3	$-\frac{\sqrt{5}}{5}$	116·6	3

10. The Six Regular Polytopes in Four Dimensions. Hence closed polytopes are possible only with 3, 4, or 5 tetrahedra at an edge, or 3 octahedra, 3 hexahedra, or 3 dodecahedra. 4 hexahedra at an edge completely fill up the space, like 4 squares in a plane; this produces in euclidean space a regular honeycomb of cubes, 4 at each edge and 8 at each vertex; this is the only regular honeycomb in euclidean space of three dimensions.

As the regular polyhedra correspond to regular networks on the sphere, so the regular polytopes correspond to regular honeycombs on the hypersphere, and regarding the hypersphere as an elliptic space of three dimensions, we have six homogeneous honeycombs in elliptic space. Their fundamental numbers are 333, 334, 335, 343, 433, 533; two are self-reciprocal: 333 and 343; the others are in reciprocal pairs: 334 and 433, 335 and 533.

The remaining four cases 534, 535, 435, 353, corresponding respectively to 4 or 5 dodecahedra, 5 hexahedra, or 3 icosahedra, at an edge, give homogeneous honeycombs in hyperbolic space.

11. We have now to determine for the regular polytopes the values of N_0, N_1, N_2, N_3. We have the configuration

N_0	2	k_1	$2\,k_1k_{12}$
$2\,k_2k_{23}$	N_1	k_1	$k_1k_2k_{12}$
$k_2k_3k_{23}$	k_3	N_2	$2\,k_2k_{12}$
$2\,k_3k_{23}$	k_3	2	N_3

where

$$k_{12} = \frac{2}{2(k_1 + k_2) - k_1k_2}, \quad k_{23} = \frac{2}{2(k_2 + k_3) - k_2k_3}.$$

Now (chap. vii (4, 1))

$$N_0k_2k_{23} = N_1, \quad N_1k_3 = N_2k_1, \quad N_2 = N_3k_2k_{12}.$$

Hence

$$N_0 : N_1 : N_2 : N_3 = k_1k_{12} : k_1k_2k_{12}k_{23} : k_2k_3k_{12}k_{23} : k_3k_{23}.$$

For the regular polyhedra in three-dimensional space the values of N_0, N_1, N_2 were determined by Euler's equation $N_0 - N_1 + N_2 = 2$. In four dimensions, however, Euler's equation is $N_0 - N_1 + N_2 - N_3 = 0$, which is homogeneous, and therefore fails to determine the absolute values of the numbers, and only their ratios are as yet determined. These ratios are shown in the following table, which gives a list of all the homogeneous honeycombs in space of three dimensions, elliptic, euclidean, and hyperbolic.

	k_1.	k_2.	k_3.	k_{12}.	k_{23}.	N_0 :	N_1 :	N_2 :	N_3
Elliptic.	3	3	3	$\frac{2}{3}$	$\frac{2}{3}$	1	2	2	1
	3	3	4	$\frac{2}{3}$	1	1	3	4	2
	4	3	3	1	$\frac{2}{3}$	2	4	3	1
	3	3	5	$\frac{2}{3}$	2	1	6	10	5
	5	3	3	2	$\frac{2}{3}$	5	10	6	1
	3	4	3	1	1	1	4	4	1
Euc.	4	3	4	1	1	1	3	3	1
Hyperbolic.	4	3	5	1	2	2	12	15	5
	5	3	4	2	1	5	15	12	2
	5	3	5	2	2	1	6	6	1
	3	5	3	2	2	1	10	10	1

To determine the absolute values of N_0, N_1, N_2, N_3 for the regular polytopes in S_4 we shall actually build up the figures. Each polytope can be identified by its fundamental numbers $k_1k_2k_3$.

12. (333). The cells are tetrahedra, 3 at each edge, 4 at each vertex.

Starting with a central tetrahedron, place a tetrahedron on each face. This covers also the edges and the vertices, since there are now 3 at each edge and 4 at each vertex. We have now 5 tetrahedra, 10 triangles, 10 edges, and 4 vertices, and all the triangles and edges are covered. The 4 new edges must therefore meet in one remaining vertex and the figure is completed (5, 10, 10, 5). It is therefore a simplex or 5-cell.

(433). The cells are cubes, 3 at each edge, 4 at each vertex.

Starting with a central cube we add a cube on each face, thus covering the edges and the vertices. Each new cube has one old face, four common to two new cubes, and one face free; four old edges, four common to three, and four common to two; four old vertices and four common to three. Hence we have added 6 cubes, $(2 + 1)6 = 18$ faces, $(\frac{4}{3} + 2)6 = 20$ edges, and $\frac{4}{3} \times 6 = 8$ vertices; and we have now 7 cubes, 24 faces, 32 edges, and 16 vertices, i.e. (7, 24, 32, 16). The outer boundary of the figure is again a cube, with 2 at each edge and 3 at each vertex, and one more cube, without any new faces, edges, or vertices, completes the figure (8, 24, 32, 16). (See Fig. 55, p. 178.) This is called the hexahedral 8-cell, or simply the 8-cell, as it is the most important figure with 8 cells in space of four dimensions. Other special isomorphic forms are the parallelotope, the analogue of the parallelepiped, and more specially the rectangular parallelotope or orthotope.

(334). This is the reciprocal of the 8-cell and can be obtained from it by taking as vertices the centres of the eight cells. The centres of the four cells at a vertex are the vertices of a tetrahedron; the centres of the 3 cells at an edge form a triangle; and the centres of two cells with common face are the ends of an edge. Hence to each vertex of the 8-cell corresponds a cell; to each edge, a face; to each face, an edge; and to each cell, a vertex. The figure is therefore a 16-cell (16, 32, 24, 8). This is the analogue of the octahedron.

(343). The cells are octahedra, 3 at each edge, 6 at each vertex.

Let us start with 6 octahedra at a vertex. The outer boundary is then of the form of a cube with a pyramid on each face, represented by the Schlegel diagram (Fig. 46). At each of the cubical edges (thick lines) there are two octahedra, and at each of the cubical vertices three; at each of the octahedral edges and vertices there is just one. Each octahedron has four free faces, and four common to two octahedra; four free edges, four common to two octahedra, and four common to three octahedra; one free vertex, four common to three octahedra, and one common to six. Hence we have so far 6, octahedra $6(4+2)=36$ faces, $6(4+2+\frac{4}{3})=44$ edges, and $6(1+\frac{4}{3}+\frac{1}{6})=15$ vertices, i.e. (6, 36, 44, 15).

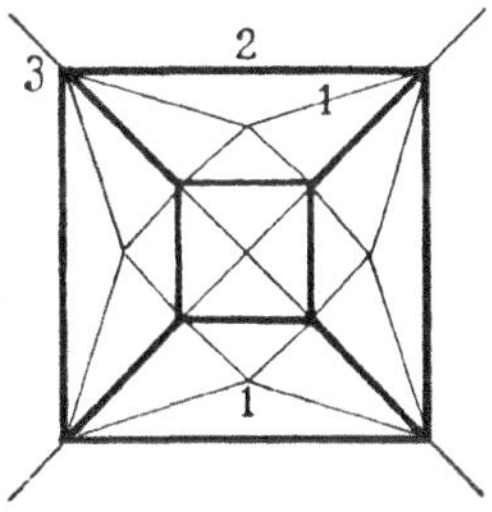

FIG. 46.

At each of the cubical edges we have to add one octahedron. We have then 6 at each cubical vertex, 5 at each octahedral vertex, and 3 at the octahedral edges, so that only the octahedral vertices remain uncovered, while new vertices, with three octahedra at each, appear in place of the old cubical vertices. The outer boundary is of the same form as before, but at cubical edges there is now just one octahedron, at octahedral edges 2, at cubical vertices 3, and at octahedral vertices 5 (Fig. 47). Each new octahedron has 2 old faces, 2 free faces, and 4 common to two octahedra; 5 old edges, 1 free edge, 4 common to two octahedra, and 2 common to three; 4 old vertices, and 2 common to three octahedra. We have therefore added 12 octahedra, $12(2 + 2) = 48$ faces, $12(1 + 2 + \frac{2}{3}) = 44$ edges, and $12 \times \frac{2}{3} = 8$ vertices, and we now have (18, 84, 88, 23).

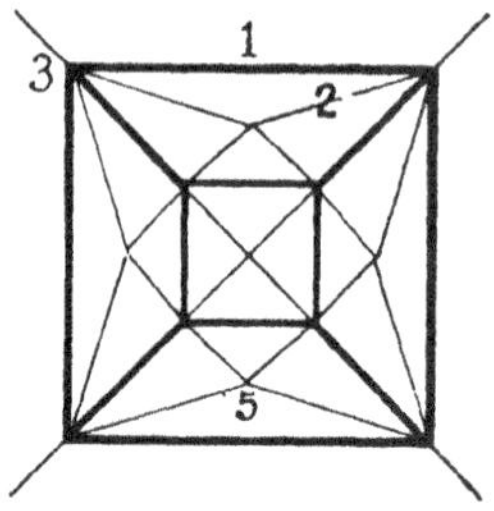

FIG. 47.

The figure is now completed by adding one octahedron at each octahedral vertex. Each new octahedron has 4 old faces, and 4 common to two octahedra; 8 old edges, and 4 common to three octahedra; 5 old vertices, and 1 common to six octahedra. Hence we have added 6 octahedra, 12 faces, 8 edges, and 1 vertex, giving (24, 96, 96, 24). This figure, the 24-cell, is self-reciprocal.

(335). The cells are tetrahedra, 5 at each edge, and 20 at each vertex.

Layer 1. Start with 20 tetrahedra at a vertex. The outer boundary is then an icosahedron; each edge has two tetrahedra, and each vertex 5, (20, 50, 42, 13). (Fig. 48).

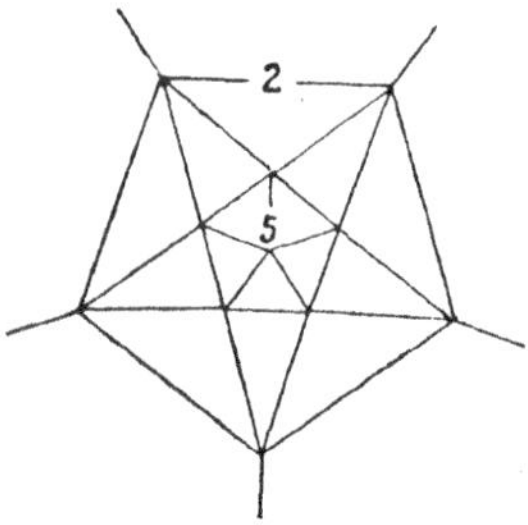

FIG. 48.

Layer 2. Adding one tetrahedron to each face, each icosahedral edge has now 4 tetrahedra, and each icosahedral vertex 10, (40, 110, 102, 33). In order to cover the edges and vertices of the first layer we require to add further one tetrahedron to each edge (70, 170, 132, 33), and 5 to each vertex (130, 290,

204, 45). To the vertices of the first layer there now correspond vertices in similar positions, and also a vertex corresponding to each of the faces of the first layer, i.e. we have both icosahedral and dodecahedral vertices. The icosahedral edges have disappeared and are replaced by dodecahedral edges. The outer boundary is now a dodecahedron with pentagonal pyramids on the faces, a "pentakis-dodecahedron" (Fig. 49). At each of the dodecahedral vertices there are 10 tetrahedra, and at each of the icosahedral vertices 5; at each of the

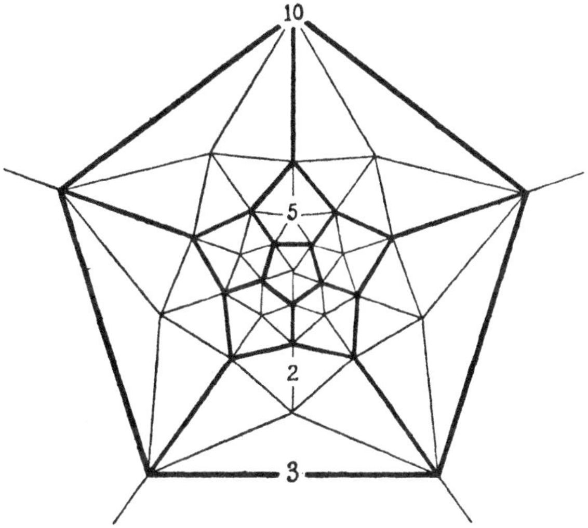

FIG. 49.

dodecahedral edges (thick lines) there are 3, and at each of the others 2.

Layer 3. To cover this layer we require now to add one tetrahedron to each face (190, 440, 324, 75), one to each of the pyramid edges (250, 560, 384, 75), one to each of the dodecahedral vertices (270, 580, 384, 75), and 5 to each of the icosahedral vertices (330, 700, 456, 87). The icosahedral vertices are now replaced by others in similar positions, while the dodecahedral vertices are replaced by triangles. The outer boundary is now a polyhedron bounded by 12 pentagons and 20 triangles, two of each at each vertex (this semi-regular polyhedron is denoted by the symbol $3_2^{20}5_2^{12}$), and a pyramid on each of the pentagons (Fig. 50). At each of the icosahedral (pyramid) vertices there are 5 tetrahedra, at each of the others 12; at each of the pyramid edges there are 2, and at each of the others 3.

Layer 4. To cover this layer we have to add one tetrahedron to each face (410, 880, 576, 107), one to each pyramid

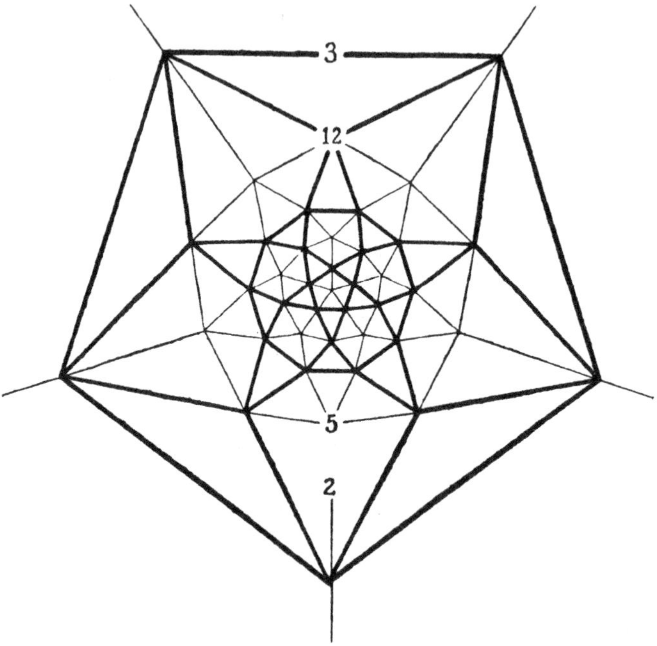

FIG. 50.

edge (470, 970, 606, 107), and 5 to each icosahedral vertex (530, 1090, 678, 119). The outer boundary is again a pentakis-dodecahedron with 16 tetrahedra at each dodecahedral vertex, 5 at each icosahedral vertex, 4 at each dodecahedral edge, and 2 at each pyramid edge (Fig. 51).

Layer 5. If we add now a tetrahedron at each dodecahedral edge (560, 1150, 708, 119), and one at each dodecahedral vertex (580, 1170, 708, 119), the boundary is again an icosahedron, with 3 tetrahedra at each edge and 15 at each vertex (Fig. 52).

The addition of one tetrahedron to each face adds 2 to each

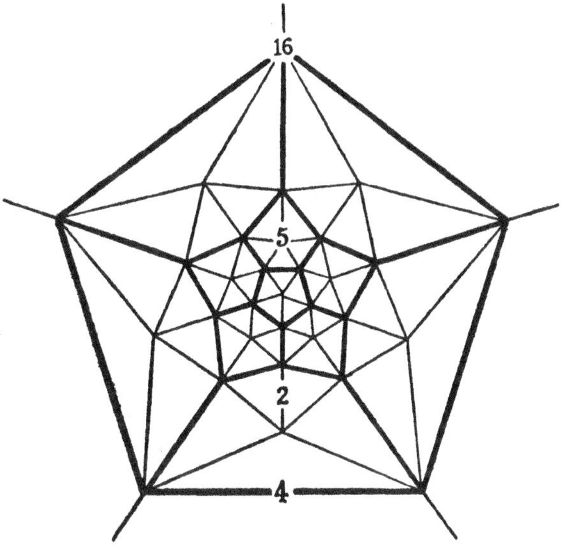

FIG. 51.

edge and 5 to each vertex, and thus completes the figure (600, 1200, 720, 120), which is thus a 600-cell.

The 120 vertices of the 600-cell have thus been divided into groups as follows:

1, 12, 20, 12, 30, 12, 20, 12, 1.

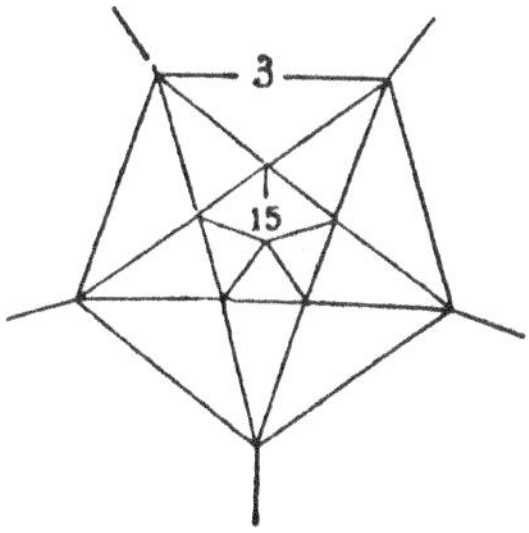

FIG. 52.

In the three-dimensional Schlegel model, with a vertex at the centre, groups of 12 vertices form regular icosahedra, groups of 20 form regular dodecahedra, and the central group forms the icosidodecahedron $3_2^{20}5_2^{12}$. A representation of the 600-cell, with a description of a model, based on this dissection of the figure, will be found in *Proc. R. Soc. Edinburgh*, vol. 34 (1914).

The 120 vertices of the 600-cell are most conveniently

denoted according to the zones in which they lie. The zones are denoted by letters A, B, C, D, E, D′, C′, B′, A′. A and A′ are single vertices. The vertices in other zones are numbered, opposite vertices of a zone being distinguished by accents. B and D consist of 12 vertices, numbered 0; 1, 2, 3, 4, 5; 1′, 2′, 3′, 4′, 5′; 0′. C consists of 20 vertices numbered 1, 2, 3, 4, 5; **1, 2, 3, 4, 5**; **1′, 2′, 3′, 4′, 5′**; 1′, 2′, 3′, 4′, 5′. In zone E, the mesial zone, the 30 vertices are numbered 1, 2, 3, 4, 5; **1, 2, 3, 4, 5**; 13, 24, 35, 41, 52; 13′ or 31, etc.; the 10 vertices of the equator of this zone being represented by the pairs of numbers which represent the vertices to which they are connected.

The edges are shown as follows:

A joined to each B.
B0 „ „ A; B 1, 2, 3, 4, 5; C 1, 2, 3, 4, 5; D 0.
1 „ „ A; B 0, 2, 3′, 4′, 5; C **1′, 3, 4**, 3, 4; D 1.
C1 „ „ B 0, 3, 4; C **1**, 2, 5; D 0, 3, 4; E **1**, 3, 4.
1 „ „ B 1′, 3, 4; C 1, **3′, 4′**; D 1′, 3, 4; E **1**, 13, 14.
D0 „ „ B 0; C 1, 2, 3, 4, 5; E 1, 2, 3, 4, 5; D′ 0.
1 „ „ B 1; C **1′, 3, 4**, 3, 4; E 1, 31, 41, **3, 4**; D′ 1.
E1 „ „ C 3, 4; D 0, 1; E 2, 5, **3, 4**; D′ 0, 1; C′ 3, 4.
1 „ „ C 1, **1**; D 3, 4; E 3, 4, 13, 14; D′ 3, 4; C′ 1, **1**.
13 „ „ C **1, 3′**; D 1′, 3; E **1, 3′**, 53, 14; D′ 1′, 3; C′ **1, 3′**.

(To complete this table: (1) perform a cyclic permutation of the numbers 12345 in each row, keeping 0 unaltered; (2) interchange accents. E′ is the same as E.)

(533). This is the reciprocal of the 600-cell, and has therefore the numbers (120, 720, 1200, 600). It is called the 120-cell, being bounded by 120 dodecahedra. In the Schlegel model with a dodecahedron at the centre we have successive zones of dodecahedra

1, 12, 20, 12, **30**, 12, 20, 12, 1.

The 600 vertices are then arranged in zones

20, 20, 30, 60, 60, 60, 20, **60**, 20, 60, 60, 60, 30, 20, 20.

Groups of 20 form the vertices of regular dodecahedra, groups of 30 form the vertices of the semi-regular icosidodecahedron, and groups of 60 form the vertices of another semi-regular polyhedron $6_2^{20}5_1^{12}$, bounded by 20 hexagons and 12 pentagons, the "icosahedron truncum."

13. As a comparison with the corresponding regular polyhedra in three dimensions we note that the vertices of the dodecahedron can be grouped in circles 5, 5, 5, 5; 1, 3, 6, 6, 3, 1; or 2, 4, 2, 4, 2, 4, 2 (Fig. 53) while those of the icosahedron

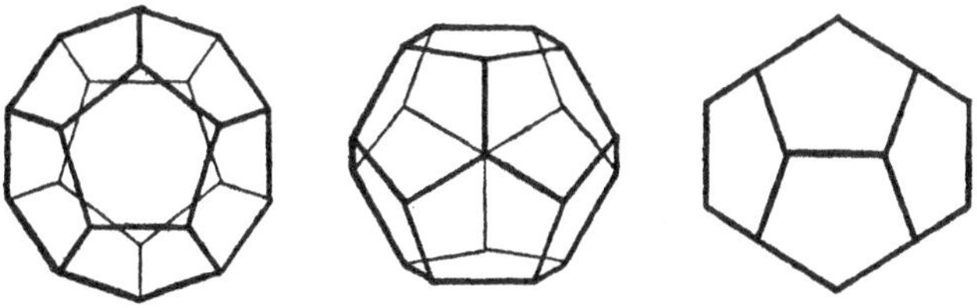

FIG. 53.—The regular dodecahedron.

fall into groups 1, 5, 5, 1; 3, 3, 3, 3; or 2, 2, 4, 2, 2 (Fig. 54).

14. Relations between the Regular Polytopes of Four Dimensions. In addition to the reciprocal relationships between the 8-cell and the 16-cell, the 120-cell and the 600-cell, there are simple relationships between the 8-cell, the 16-cell,

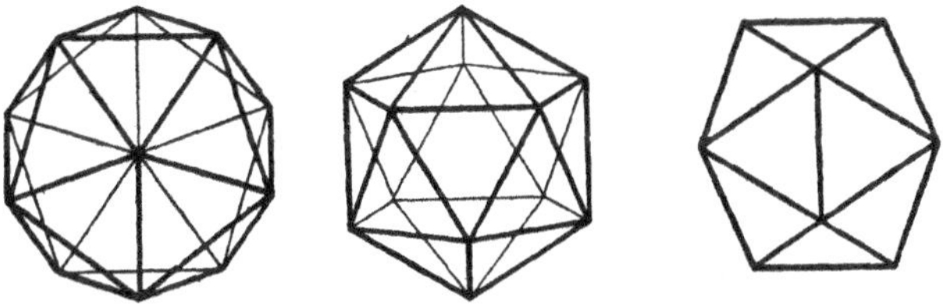

FIG. 54.—The regular icosahedron.

and the 24-cell analogous to the relationship of a tetrahedron to a cube.

The vertices of a regular 8-cell fall into two groups which are vertices of two regular 16-cells.

Denote the vertices of one cube of the 8-cell by ABCDPQRS (Fig. 55), ABCD being a regular tetrahedron and P, Q, R, S the opposite vertices of the cube; and let A′, B′, etc., be the opposite vertices of the 8-cell. The 8 cells are then

A B C D P Q R S	A′B′C′D′P′Q′R′S′
A′B C D′S R′Q′P	A B′C′D S′R Q P′
A B′C D′R′S P′Q	A′B C′D R S′P Q′
A B C′D′Q′P′S R	A′B′C D Q P S′R′.

The lines joining pairs of the 8 vertices A, B, C, D, A′, B′, C′, D′, with the exception of pairs of opposite vertices AA′, etc.,

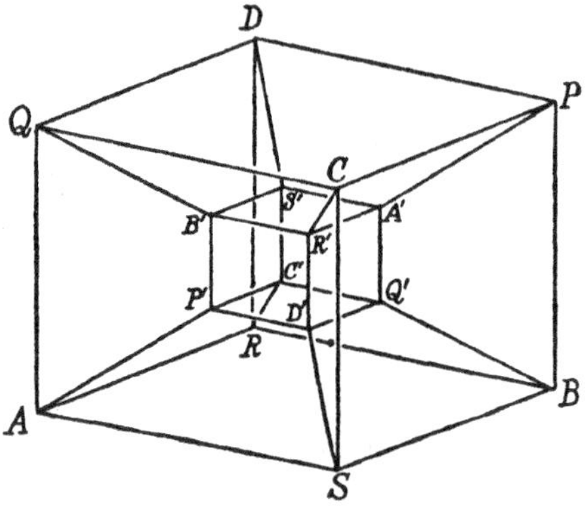

FIG. 55.

are diagonals of squares and are therefore all equal. Hence there are 16 regular tetrahedra:

A B C D	A′B′C′D′	A B C D′	A′B′C′D
A′B C D	A B′C′D′	A′B C D′	A B′C′D
A B′C D	A′B C′D′	A B′C D′	A′B C′D
A B C′D	A′B′C D′	A B C′D′	A′B′C D

which form a regular 16-cell. Similarly a 16-cell is formed with the 8 vertices PQRSP′Q′R′S′.

The vertices of a regular 24-*cell can be divided into three separate groups, each forming the vertices of a regular* 16-*cell.*

Referring to the construction of the 24-cell (p. 171) we see that if we omit one vertex O (the central vertex, say, in the Schlegel model), the opposite vertex O′, and the six vertices opposite to O of the six octahedra which meet at O, which are also the vertices opposite to O′ of the six which meet at O′, we are left with 16 vertices which are those of a regular 8-cell. The vertices omitted are those of a regular 16-cell. Considering the hyperspherical network corresponding to the 24-cell, i.e. a division of elliptic three-dimensional space into 24 octahedra, we may derive this division as follows. First divide elliptic space into 8 regular hexahedra, 3 at each edge. The centres of the cells are then vertices of a regular tetrahedral 16-cellular honeycomb. Each hexahedron is divided into six quadrilateral pyramids, and two quadrilateral pyramids on opposite sides of a face of a hexahedral cell together form a regular octahedron (Fig. 56). We have therefore 6 octahedra at the centre of each hexahedral cell; and since there are 4 hexahedra at a vertex there are 12 pyramids and therefore 6 octahedra also at each hexahedral vertex. Also since there are three hexahedra at each edge there are three octahedra at each hexahedral edge, and there are three also at each of the other edges. Half of each octahedron is one-sixth of a hexahedron, hence there are $3 \times 8 = 24$ octahedral cells.

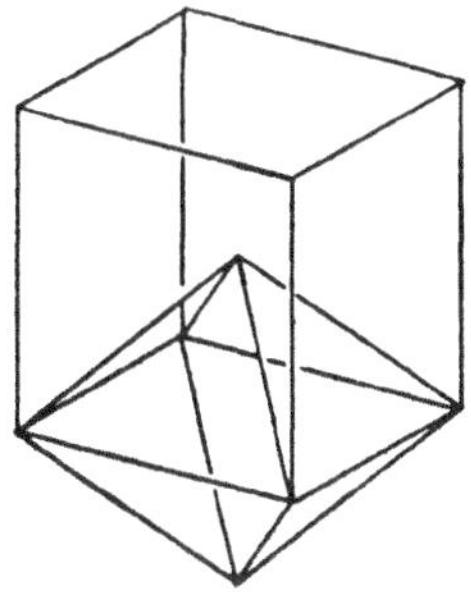

FIG. 56.

Thus the 16 vertices of the 8-cell and the centres of the 8-cells (which are also vertices of a 16-cell) are vertices of a 24-cell; and since the vertices of the 8-cell can be further divided into two groups of vertices of 16-cells, the 24 vertices of the 24-cell can be divided into three groups of vertices of 16-cells.

15. The Regular Polytopes Referred to Rectangular Axes. We shall consider this problem first for the three-dimensional figures.

(1) *The Cube.* The simplest rectangular axes are those through the centre parallel to the edges (Fig. 57). The length of the edges being 2, the co-ordinates of the vertices are:

A(1, − 1, − 1),	A′(− 1, 1, 1),
B(− 1, 1, − 1),	B′(1, − 1, 1),
C(− 1, − 1, 1),	C′(1, 1, − 1),
D(1, 1, 1),	D′(− 1, − 1, − 1).

(2) *The Tetrahedron.* In the same figure ABCD is a regular tetrahedron. The length of the edge $= 2\sqrt{2}$.

(3) *The Octahedron.* Take as axes the three body diagonals (Fig. 58). Then the co-ordinates are

$$\begin{array}{ll} X(1, 0, 0), & X'(-1, 0, 0), \\ Y(0, 1, 0), & Y'(0, -1, 0), \\ Z(0, 0, 1), & Z'(0, 0, -1), \end{array}$$

the length of the edge being $\sqrt{2}$.

(4) *The Dodecahedron.* Consider three contiguous faces of a dodecahedron, DRNA'P, DPLB'Q, DQMC'R, meeting at D (Fig. 59), and draw the diagonals DA', DB', DC'. Now DA' || RN, and DB' || PL, but the edges PL and RN are orthogonal, hence DA' ⊥ DB', and similarly ⊥ DC' also; and

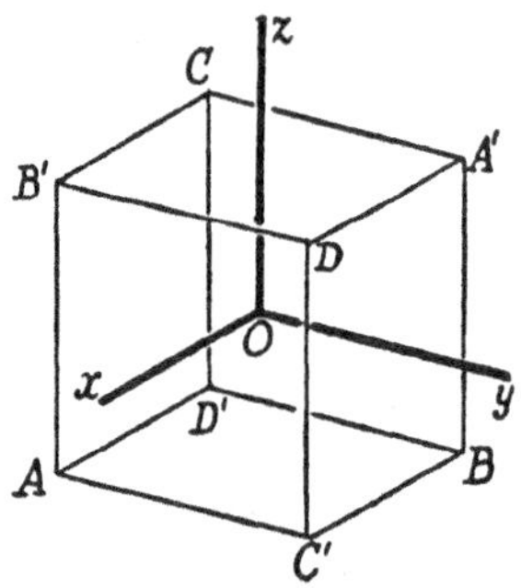

FIG. 57.

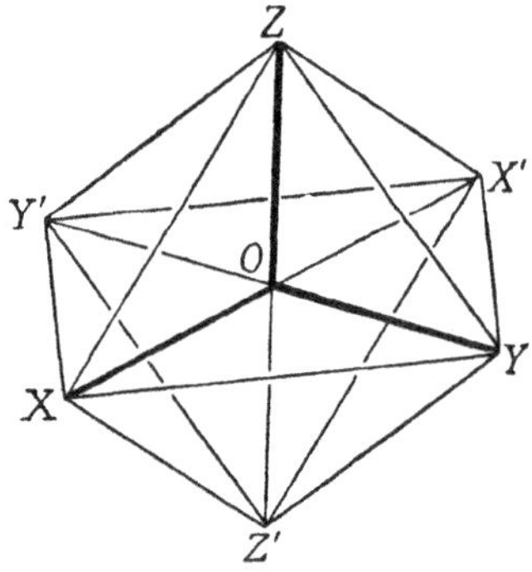

FIG. 58.

since these diagonals are equal they form the edges of a cube. The vertices of this cube are ABCDA'B'C'D'. As each face has 5 diagonals, there are 5 ways of choosing 8 vertices of a regular dodecahedron to form the vertices of a cube.

We can now choose a set of rectangular axes through the centre of the dodecahedron parallel to the edges of the cube. Taking the length of the edge of the cube = $2c$, and the edge of the dodecahedron = $2a$, we have (Fig. 60)

$$\begin{aligned} c &= 2a \cos 36^\circ = \tfrac{1}{2}(\sqrt{5} + 1)a, \\ a &= \tfrac{1}{2}(\sqrt{5} - 1)c, \\ MN &= c - a = \tfrac{1}{2}(\sqrt{5} - 1)a, \\ PN &= 2a \sin 36^\circ = \tfrac{1}{2}\sqrt{(10 - 2\sqrt{5})}a. \end{aligned}$$

Hence $PM^2 = PN^2 - MN^2 = \frac{1}{4}(10 - 2\sqrt{5} - 6 + 2\sqrt{5})a^2 = a^2$,
i.e. $PM = a = \frac{1}{2}(\sqrt{5} - 1)c$.

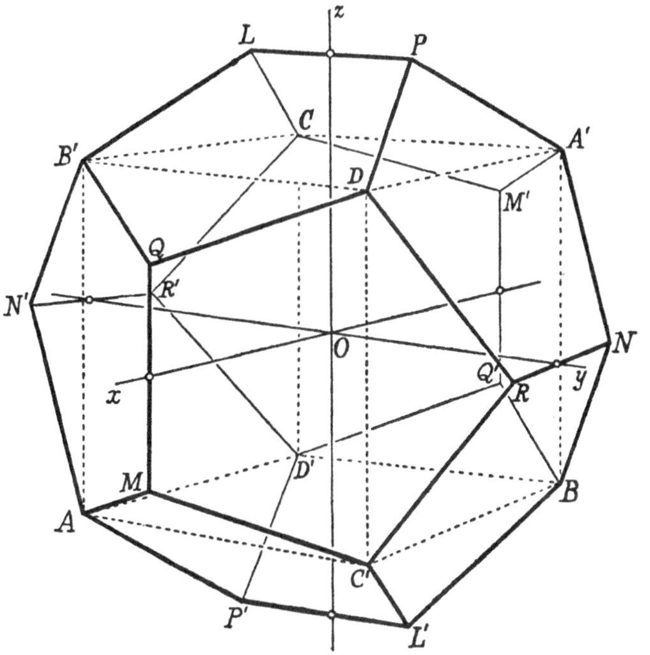

FIG. 59.

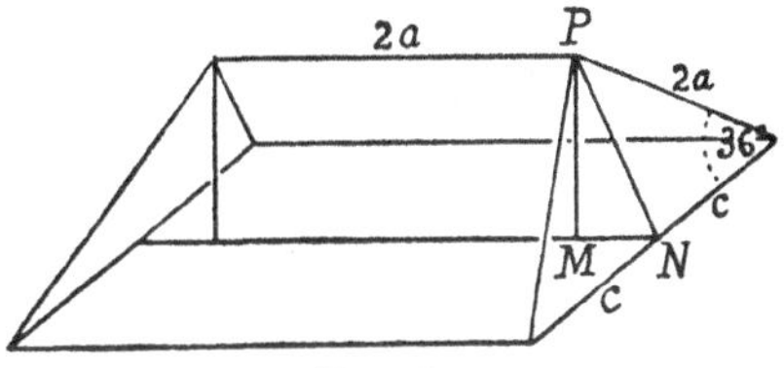

FIG. 60.

The co-ordinates of the vertices of the dodecahedron are then

A, B, C, D, A′, B′, C′, D′	$\pm c$,	$\pm c$,	$\pm c$,
P, L, P′, L′	0,	$\pm a$,	$\pm (c + a)$,
Q, M, Q′, M′	$\pm (c + a)$	0,	$\pm a$,
R, N, R′, N′	$\pm a$,	$\pm (c + a)$,	0;

or, if the edge of the dodecahedron is unity, and e stands for $\surd 5$, we have

8 vertices	$\pm \frac{1}{4}(e+1)$,	$\pm \frac{1}{4}(e+1)$,	$\pm \frac{1}{4}(e+1)$,
4 vertices	0,	$\pm \frac{1}{2}$,	$\pm \frac{1}{4}(e+3)$,
4 vertices	$\pm \frac{1}{4}(e+3)$,	0,	$\pm \frac{1}{2}$,
4 vertices	$\pm \frac{1}{2}$,	$\pm \frac{1}{4}(e+3)$,	0.

(5) *The Icosahedron.* The centroids of the faces of the dodecahedron are vertices of a regular icosahedron. Taking the face DRNA′P (Fig. 59) whose vertices have co-ordinates

$$\begin{array}{ccc} c, & c, & c, \\ a, & c+a, & 0, \\ -a, & c+a, & 0, \\ -c, & c, & c, \\ 0, & a, & c+a, \end{array}$$

the co-ordinates of the centroid are 0, $\frac{1}{5}(4c+3a)$, $\frac{1}{5}(3c+a)$, and for the centroid of the face BNRC′L′: 0, $\frac{1}{5}(4c+3a)$, $-\frac{1}{5}(3c+a)$. The distance between these adjacent vertices is $\frac{2}{5}(3c+a) = \frac{1}{5}(3\surd 5+5)a$. Hence, if the edge of the icosahedron is unity, the co-ordinates of the vertices are

$$\begin{array}{ccc} 0, & \pm \frac{1}{4}(e+1), & \pm \frac{1}{2}, \\ \pm \frac{1}{2}, & 0, & \pm \frac{1}{4}(e+1), \\ \pm \frac{1}{4}(e+1), & \pm \frac{1}{2}, & 0. \end{array}$$

16. The Regular Polytopes of Four Dimensions Referred to Rectangular Axes. (1) *The 8-Cell.* Taking the origin at the centre and axes parallel to the edges, the co-ordinates of the 16 vertices are (see Fig. 55, p. 178)

$$\begin{array}{llll} P(\;\;1,-1,-1,\;\;1), & A(-1,\;\;1,\;\;1,\;\;1), \\ Q(-1,\;\;1,-1,\;\;1), & B(\;\;1,-1,\;\;1,\;\;1), \\ R(-1,-1,\;\;1,\;\;1), & C(\;\;1,\;\;1,-1,\;\;1), \\ S(\;\;1,\;\;1,\;\;1,\;\;1), & D(-1,-1,-1,\;\;1), \end{array}$$

the other 8 vertices A′, . . ., P′, . . . having the co-ordinates all reversed in sign. The cells are formed of eight vertices with different letters, all unaccented, all accented, or two of ABCD accented, say A′B′CD, and the two corresponding letters of PQRS unaccented, PQR′S′. Pairs of vertices with the same letter, accented and unaccented, as AA′, having all the co-ordinates different, form 8 body diagonals. Pairs of ABCD or PQRS, each with or without accent, having two co-ordinates different, form 48 diagonals of squares. Pairs such as AP, AQ′, AR′, AS′, (not AP′), having three co-ordinates different, form 32 diagonals of cubes. Pairs such as AP′, AQ, AR, AS, having one co-ordinate different, form the 32 edges. The length of the edge = 2.

(2) *The* 16-*Cell.* Taking for axes the four body-diagonals, the co-ordinates of the vertices are

$$\begin{array}{cccc} (\pm 1, & 0, & 0, & 0), \\ (\ 0, & \pm 1, & 0, & 0), \\ (\ 0, & 0, & \pm 1, & 0), \\ (\ 0, & 0, & 0, & \pm 1), \end{array}$$

the length of the edge being $\sqrt{2}$.

The vertices of the 16-cell may also be taken as half the vertices of the 8-cell, say ABCDA′B′C′D′ or PQRSP′Q′R′S′. Its edges are diagonals of squares, of length $2\sqrt{2}$.

(3) *The* 8-*Cell.* Another set of axes for the 8-cell may be obtained from this last representation by taking the centroids of the cells of the 16-cell. The centroids are: for ABCD, (0, 0, 0, 1); for ABCD′, ($\frac{1}{2}$, $\frac{1}{2}$, $\frac{1}{2}$, $\frac{1}{2}$). Hence we obtain as vertices of the 8-cell

$$\begin{array}{llll} P(\ 1, -1, -1, 1), & P'(-1, \ 1, \ 1, -1), & F, F'(\pm 2, \ 0, \ 0, \ 0), \\ Q(-1, \ 1, -1, 1), & Q'(\ 1, -1, \ 1, -1), & G, G'(\ 0, \pm 2, \ 0, \ 0), \\ R(-1, -1, \ 1, 1), & R'(\ 1, \ 1, -1, -1), & H, H'(\ 0, \ 0, \pm 2, \ 0), \\ S(\ 1, \ 1, \ 1, 1), & S'(-1, -1, -1, -1), & K, K'(\ 0, \ 0, \ 0, \pm 2), \end{array}$$

the length of the edge being 2. The vertices PP′QQ′RR′SS′ form one 16-cell, and the vertices FF′GG′HH′KK′ another 16-cell. The 32 edges are FP, FQ′, FR′, FS, KP, KS; the others being obtained from these by cyclic permutations (PQR)(FGH) and interchanging accents.

(4) *The* 24-*Cell.* The vertices of a 24-cell are the vertices and cell-centres of an 8-cell. Hence we can represent them by ABCDA′B′C′D′, PQRSP′Q′R′S′, FGHKF′G′H′K′, with co-ordinates (± 1, ± 1, ± 1, ± 1), (± 2, 0, 0, 0), (0, ± 2, 0, 0), (0, 0, ± 2, 0), (0, 0, 0, ± 2). The edges are FA′, FB, FC, FD′; FP, FQ′, FR′, FS; PA′, PB, PC, PD; KA, KB, KC,

KD; KP, KQ, KR, KS; SA, SB, SC, SD′, and others obtained by the simultaneous cyclic permutations (ABC) (PQR) (FGH) and interchange of accents. The length of the edge = 2.

(5) *The* 5-*Cell.* Take one of the tetrahedra as (xyz)-co-ordinate-flat, the co-ordinates of its four vertices being $(-1, 1, 1, 0)$, $(1, -1, 1, 0)$, $(1, 1, -1, 0)$, $(-1, -1, -1, 0)$. The line joining the centroid C of this tetrahedron to the opposite vertex of the 5-cell is the axis of w, and the co-ordinates of the 5th vertex are $(0, 0, 0, c)$. The edges are all equal if $8 = 3 + c^2$, hence $c = \sqrt{5} = e$. The co-ordinates of the centroid of the 5-cell are now $(0, 0, 0, \frac{1}{5}e)$, so that transforming to parallel axes through the centroid we obtain the vertices

$$(-1, 1, 1, -\tfrac{1}{5}e),\ (1, -1, 1, -\tfrac{1}{5}e),\ (1, 1, -1, -\tfrac{1}{5}e),$$
$$(-1, -1, -1, -\tfrac{1}{5}e),\quad (0, 0, 0, \tfrac{4}{5}e).$$

The length of the edge $= 2\sqrt{2}$.

(6) *The* 600-*Cell.* The vertices of the 600-cell have been grouped in zones of 1, 12, 20, 12, 30, 12, 20, 12, 1. The central zone consists of the semi-regular icosidodecahedron whose vertices are the mid-points of the edges of a regular dodecahedron. Referring this to rectangular axes we find the co-ordinates of its 30 vertices:

$$\begin{array}{ccc}
\pm \tfrac{1}{2}(e+1), & 0, & 0, \\
0, & \pm \tfrac{1}{2}(e+1), & 0, \\
0, & 0, & \pm \tfrac{1}{2}(e+1),
\end{array}$$

$$\begin{array}{ccc}
\pm \tfrac{1}{4}(e+3), & \pm \tfrac{1}{2}, & \pm \tfrac{1}{4}(e+1), \\
\pm \tfrac{1}{4}(e+1), & \pm \tfrac{1}{4}(e+3), & \pm \tfrac{1}{2}, \\
\pm \tfrac{1}{2}, & \pm \tfrac{1}{4}(e+1), & \pm \tfrac{1}{4}(e+3),
\end{array}$$

the length of each edge being unity, and the radius of the circumscribed sphere $\frac{1}{2}(e + 1)$. For each of these vertices the fourth co-ordinate $w = 0$, and the co-ordinates of the two vertices at each end of the series of zones are $\{0, 0, 0, \pm \frac{1}{2}(e + 1)\}$. If we interchange the axes of x and y and at the same time those of z and w we get further vertices, and thus obtain the 104 vertices:

$$\begin{array}{cccc}
\pm\frac{1}{2}(e+1), & 0, & 0, & 0, \\
0, & \pm\frac{1}{2}(e+1), & 0, & 0, \\
0, & 0, & \pm\frac{1}{2}(e+1), & 0, \\
0, & 0, & 0, & \pm\frac{1}{2}(e+1);
\end{array}$$

$$\begin{array}{cccc}
0, & \pm\frac{1}{4}(e+1), & \pm\frac{1}{2}, & \pm\frac{1}{4}(e+3), \\
0, & \pm\frac{1}{2}, & \pm\frac{1}{4}(e+3), & \pm\frac{1}{4}(e+1), \\
0, & \pm\frac{1}{4}(e+3), & \pm\frac{1}{4}(e+1), & \pm\frac{1}{2};
\end{array}$$

$$\begin{array}{cccc}
\pm\frac{1}{4}(e+1), & 0, & \pm\frac{1}{4}(e+3), & \pm\frac{1}{2}, \\
\pm\frac{1}{2}, & 0, & \pm\frac{1}{4}(e+1), & \pm\frac{1}{4}(e+3), \\
\pm\frac{1}{4}(e+3), & 0, & \pm\frac{1}{2}, & \pm\frac{1}{4}(e+1);
\end{array}$$

$$\begin{array}{cccc}
\pm\frac{1}{2}, & \pm\frac{1}{4}(e+3), & 0, & \pm\frac{1}{4}(e+1), \\
\pm\frac{1}{4}(e+3), & \pm\frac{1}{4}(e+1), & 0, & \pm\frac{1}{2}, \\
\pm\frac{1}{4}(e+1), & \pm\frac{1}{2}, & 0, & \pm\frac{1}{4}(e+3);
\end{array}$$

$$\begin{array}{cccc}
\pm\frac{1}{4}(e+3), & \pm\frac{1}{2}, & \pm\frac{1}{4}(e+1), & 0, \\
\pm\frac{1}{4}(e+1), & \pm\frac{1}{4}(e+3), & \pm\frac{1}{2}, & 0, \\
\pm\frac{1}{2}, & \pm\frac{1}{4}(e+1), & \pm\frac{1}{4}(e+3), & 0.
\end{array}$$

We have now obtained zones of vertices at distances $\frac{1}{2}(e+1)$, $\frac{1}{4}(e+3)$, $\frac{1}{4}(e+1)$, $\frac{1}{2}$, and 0, from the central zone, and the numbers in these zones are 1, 12, 12, 12, 30. Hence there are 8 more vertices with $w = \frac{1}{4}(e+1)$. As from symmetry there must also be 8 more with $x = \frac{1}{4}(e+1)$, etc., the additional 16 vertices can only be

$$\pm\frac{1}{4}(e+1),\ \pm\frac{1}{4}(e+1),\ \pm\frac{1}{4}(e+1),\ \pm\frac{1}{4}(e+1).$$

We verify that these vertices also are at a distance $\frac{1}{2}(e+1)$ from the centre.

The following table identifies the co-ordinates with the vertices according to the notation of § 12. The edge is taken equal to 4:

	x.	*y*.	*z*.	*w*.
A	0	0	0	$2(e+1)$
B 0, 1	0	± 2	$e+1$	$e+3$
2, 5′	$e+1$	0	± 2	$e+3$
3, 4	± 2	$e+1$	0	$e+3$
C 1, 1	0	$e+3$	± 2	$e+1$
4, 3	± 2	0	$e+3$	$e+1$
5, 2′	$e+3$	± 2	0	$e+1$
5, 3′, 4, 2′	$e+1$	$\pm(e+1)$	$\pm(e+1)$	$e+1$
D 0, 1	0	$\pm(e+1)$	$e+3$	2
2, 5′	$e+3$	0	$\pm(e+1)$	2
3, 4	$\pm(e+1)$	$e+3$	0	2
E 52	$2(e+1)$	0	0	0
1	0	$2(e+1)$	0	0
1	0	0	$2(e+1)$	0
3, 13, 41, 4′	2	$\pm(e+3)$	$\pm(e+1)$	0
2, 3′, 5, 4′	$\pm(e+1)$	2	$\pm(e+3)$	0
5, 42, 2, 35	$\pm(e+3)$	$\pm(e+1)$	2	0

The order of the combinations of sign, corresponding to the vertices going from left to right in a row is + +, + −, − +, − − . To complete the table change the accents in the numbers and reverse the signs of x, y, z; accent the letters and change the sign of w.

17. The Homogeneous Honeycombs in n Dimensions. We have now to complete the problem of determining the homogeneous polytopes in space of n dimensions, which is part of the somewhat wider problem of determining the homogeneous honeycombs in space of $n-1$ dimensions. Considering the problem from the second point of view we have to divide space of $n-1$ dimensions into cells which are all homogeneous closed polytopes of $n-1$ dimensions of the same form, i.e., homogeneous honeycombs of elliptic space of $n-2$ dimensions; and further, if a small hypersphere is described round each vertex the honeycomb into which its surface is divided is also a homogeneous honeycomb of elliptic space of $n-2$ dimensions.

Consider any boundary A_{p+1} of $p+1$ dimensions, and in it any boundary A_q of q dimensions ($q \gtrless p$). In A_{p+1} take any $(p-q+1)$-flat S_{p-q+1} cutting A_q in a point O, and construct a small hypersphere with centre O. Boundaries of $q + r$ dimensions through A_q are cut by S_{p-q+1} in r-flats through O, and these cut the hypersphere in $(r - 1)$-dimensional regions; in particular the $(q + 1)$-boundaries through A_q cut the hypersphere in points, the $(q + 2)$-boundaries in arcs of great circles, and so on. Thus the figure formed on the hypersphere is a homogeneous elliptic honeycomb of $p - q$ dimensions, and if ${}_aN'_{bc}$ are its configurational numbers

$$ {}_aN'_{bc} = {}_{a+q}N_{b+q,\, c+q}. $$

In particular for $q = p - 1$ we have an elliptic honeycomb of 1 dimension, i.e. a polygon, and

$$ {}_{p+1}N_{p,\, p-1} = {}_2N'_{10} = 2; $$

for $q = p - 2$ we have a 2-dimensional honeycomb, and

$$ {}_{p+1}N_{p,\, p-2} = {}_3N'_{20} = {}_3N'_{10} = {}_{p+1}N_{p-1,\, p-2}. $$

The numbers

$$ {}_{p+1}N_{p,\, p-2} = {}_{p+1}N_{p-1,\, p-2} = k_p \quad (p = 1, 2, \ldots, n - 1) $$

are called the *fundamental numbers* of the honeycomb. The notation is extended to hold for $p = 1$, with the understanding that ${}_rN_{p,\,-1}$ means the total number of p-boundaries in an r-boundary, $= N_{pr}$.

Now let ${}_pF_{qr}$ be the configurational numbers of the face-constituents or cells of the complete honeycomb of $n - 1$ dimensions, then

$$ {}_{p+1}F_{p,\, p-2} = {}_{p+1}N_{p,\, p-2} = k_p. $$

Hence the fundamental numbers of the face-constituents are $k_1, k_2, \ldots, k_{n-2}$.

Let ${}_pV_{qr}$ be the configurational numbers of the vertex-constituents or assemblage of elements through a vertex, then

$$ {}_{p+1}V_{p,\, p-2} = {}_{p+2}N_{p+1,\, p-1} = k_{p+1}, $$

Hence the fundamental numbers for the vertex-constituents are $k_2, k_3, \ldots, k_{n-1}$.

18. The Eleven Homogeneous Honeycombs in Four

Dimensions. For a finite polytope in n dimensions with closed cells both the face-constituents and the vertex-constituents must be elliptic honeycombs or closed polytopes. For finite homogeneous polytopes in four dimensions or elliptic honeycombs in three dimensions we have seen that the only sets of fundamental numbers are

333, 334, 335, 343, 433, 533.

Hence for homogeneous polytopes in five dimensions or honeycombs in four dimensions with closed cells the only sets of fundamental numbers possible are

3333, 3334, 3335, 3343, 3433, 4333,
4334, 4335, 5333, 5334, 5335.

19. We have now to discriminate as before whether these are elliptic, euclidean, or hyperbolic, and as the same problem will recur successively in passing from one dimension to the next we shall investigate a general criterion depending on the dihedral angles.

Consider a regular honeycomb in space of n dimensions (elliptic, euclidean, or hyperbolic), with the fundamental numbers $k_1, k_2, \ldots, k_n$. Let C_0 be any vertex (Fig. 44, p. 168), C_1 the mid-point of an edge through C_0, C_2 the centre of a two-dimensional boundary (polygon) through C_0C_1, and in general C_r the centre of an r-boundary through $C_0C_1 \ldots C_{r-1}$. $C_pC_qC_r$ $(p < q < r)$ is always a right-angled triangle with right angle at C_q.

The angle $C_0C_2C_1$ is half the angle at the centre of a plane face subtended by an edge and is equal to π/k_1. The angle $C_1C_0C_2$ is half the angle between two adjacent edges, $= \theta_1$, say; $C_2C_1C_3$ is half the dihedral angle between two adjacent plane faces, $= \theta_2$, say; in general let $C_pC_{p-1}C_{p+1}$, half the dihedral angle between two adjacent p-boundaries, be denoted by θ_p. Since there are k_n boundaries of $n - 1$ dimensions at each $(n - 2)$-boundary the angle $C_{n-1}C_{n-2}C_n = \theta_{n-1} = \pi/k_n$.

In the 3-dimensional boundary $C_0C_1C_2C_3$ draw a small sphere with centre C_0. This is cut by the lines and planes through C_0 in a regular spherical polygon with sides $2\theta_1$ and angles $2\theta_2$, and the angle subtended at the centre of the polygon by half the side $= \pi/k_2$. Hence by spherical trigonometry

$$\cos \pi/k_2 = \sin \theta_2 \cos \theta_1.$$

In the 4-boundary $C_0 \ldots C_4$ take a hyperplane α perpendicular to C_0C_1 at C_1, and in this hyperplane draw a small sphere with centre C_1. This is cut by the lines and planes in which α cuts the planes and hyperplanes through C_0C_1 in a regular spherical polygon with sides $2\theta_2$ and angles $2\theta_3$, and the angle subtended at the centre of the polygon by half the side $= \pi/k_3$. Hence

$$\cos \pi/k_3 = \sin \theta_3 \cos \theta_2.$$

Proceeding in this way we have the recurring formula

$$\cos \pi/k_r = \sin \theta_r \cos \theta_{r-1} \qquad (r = 2, 3, \ldots, n-1)$$

and as $\theta_{n-1} = \pi/k_n$ this formula determines in succession $\theta_{n-2}, \theta_{n-3}, \ldots, \theta_1$.

As the formulæ of spherical trigonometry are the same for elliptic, euclidean, or hyperbolic geometry, the geometry at an angle being always elliptic, the foregoing investigation holds whether space is elliptic, euclidean, or hyperbolic. To discriminate this we take now the angles in a plane face, and consider the right-angled triangle $C_0C_1C_2$. In euclidean geometry $\theta_1 + \pi/k_1 = \pi/2$, in elliptic geometry $\theta_1 > \frac{\pi}{2} - \frac{\pi}{k_1}$, and in hyperbolic geometry it is less.

20. The Three Regular Polytopes in Five or more Dimensions. Now for a honeycomb in 4 dimensions

$$\cos \pi/k_3 = \sin \theta_3 \cos \theta_2 = \sin \pi/k_4 \cos \theta_2,$$

and

$$\cos \pi/k_2 = \sin \theta_2 \cos \theta_1.$$

Hence

$$\frac{\cos^2 \pi/k_3}{\sin^2 \pi/k_4} + \frac{\cos^2 \pi/k_2}{\cos^2 \theta_1} = 1,$$

and the honeycomb will be elliptic, euclidean, or hyperbolic, according as

$$\frac{\cos^2 \pi/k_3}{\sin^2 \pi/k_4} + \frac{\cos^2 \pi/k_2}{\sin^2 \pi/k_1} <, =, \text{ or } > 1.$$

Applying this to the eleven sets of values of k_1, k_2, k_3, k_4 we find that

3333, 3334, 4333 are elliptic,
3343, 3433, 4334 are euclidean,
3335, 5333, 4335, 5334, 5335 are hyperbolic.

Hence in space of five dimensions there are just three regular polytopes. 3333 is a regular simplex, 4333 is a regular orthotope, and 3334 is the reciprocal of the orthotope. We see also that euclidean space of four dimensions can be divided into three regular honeycombs: 3343 has 16-cells as its cells, 3433 (the reciprocal of the former) has 24-cells as its cells, and 4334 has orthotopes and is the analogue of the square network in two dimensions and the cubical honeycomb in three dimensions.

21. For a honeycomb in five dimensions the only sets of values for the fundamental numbers, since $k_1k_2k_3k_4$ and $k_2k_3k_4k_5$ must be both elliptic, are

33333, 33334, 43333, 43334.

Clearly the first gives a simplex in S_6, the third an orthotope, and the second the reciprocal of the orthotope; while 43334 is a honeycomb in euclidean space of five dimensions with orthotopes for cells.

For space of higher dimensions these series persist and are the only ones possible. Recapitulating, in a plane there are an unlimited number of regular polygons and 3 regular networks, in space of three dimensions there are 5 regular polyhedra and one regular honeycomb, in S_4 there are 6 regular polytopes and three regular honeycombs, in space of more than four dimensions there are just 3 regular polytopes and one regular honeycomb. The following is a complete table of the fundamental numbers of all the honeycombs with closed cells:

Dimensions.	2.	3.	4.	$n > 4$.
Elliptic .	33 34 43 35 53	333 334 433 335 533 343	3333 3334 4333	33 . . . 33 33 . . . 34 43 . . . 33
Euclidean .	44 36 63	434	4334 3433 3343	43 . . . 34
Hyperbolic .	k_1k_2 $2(k_1+k_2) < k_1k_2$	435 534 535 353	4335 5334 5335 3335 5333	

REFERENCES

SCHLÄFLI, L. (see chap. viii.). (The second part of this paper, published in 1860, gives the first enumeration of the regular polytopes in space of *n* dimensions, though priority is often ascribed to Stringham.)

SCHOUTE, P. H. Regelmässige Schnitte und Projectionen des Achtzelles, Sechzehnzelles und Vierundzwanzigzelles im vierdimensionalen Raume. Amsterdam, Verh. K. Akad. Wet., II., No. 2 (1893).

— . . . des 120-zelles und 600-zelles. . . . *Ibid.* No. 7, and IX., No. 4 (1907).

SOMMERVILLE, D. M. Y. The regular divisions of space of *n* dimensions and their metrical constants. Palermo, Rend. Circ. mat., **48** (1924), 9-22.

— Description of a projection-model of the 600-cell in space of four dimensions. Proc. R. Soc. Edinburgh, **34** (1914), 253-258 (1 plate).

STRINGHAM, W. I. Regular figures in *n*-dimensional space. Amer. J. Math., **3** (1880), 1-14.

(A set of projection-models of the regular four-dimensional figures, prepared by V. Schlegel, was exhibited at the Nürnberg meeting of the Deutsche Math. Vereinigung, 1892, and were afterwards on sale by M. Schilling. A set of these models, from Professor Steggall's collection in Dundee, was shown at the exhibition in connection with the Napier Tercentenary at Edinburgh, 1914. These are described in the following.)

DYCK, W. Katalog mathematischer Modelle. München, 1892. Deutsche Math. Ver. Nürnberger Ausstellung, p. 253.

SCHILLING, M. Catalog math. Modelle. Leipzig, 7. Aufl., 1911.

STEGGALL, J. E. A. Napier Tercentenary Celebration Handbook, ed. E. M. Horsburgh. R. Soc. Edinburgh, or London: Bell, 1914. p. 319.

INDEX

(*The references are to the pages*)

A CATALOG OF SELECTED

DOVER BOOKS

IN SCIENCE AND MATHEMATICS

Mathematics-Bestsellers

HANDBOOK OF MATHEMATICAL FUNCTIONS: with Formulas, Graphs, and Mathematical Tables, Edited by Milton Abramowitz and Irene A. Stegun. A classic resource for working with special functions, standard trig, and exponential logarithmic definitions and extensions, it features 29 sets of tables, some to as high as 20 places. 1046pp. 8 x 10 1/2. 0-486-61272-4

ABSTRACT AND CONCRETE CATEGORIES: The Joy of Cats, Jiri Adamek, Horst Herrlich, and George E. Strecker. This up-to-date introductory treatment employs category theory to explore the theory of structures. Its unique approach stresses concrete categories and presents a systematic view of factorization structures. Numerous examples. 1990 edition, updated 2004. 528pp. 6 1/8 x 9 1/4. 0-486-46934-4

MATHEMATICS: Its Content, Methods and Meaning, A. D. Aleksandrov, A. N. Kolmogorov, and M. A. Lavrent'ev. Major survey offers comprehensive, coherent discussions of analytic geometry, algebra, differential equations, calculus of variations, functions of a complex variable, prime numbers, linear and non-Euclidean geometry, topology, functional analysis, more. 1963 edition. 1120pp. 5 3/8 x 8 1/2. 0-486-40916-3

INTRODUCTION TO VECTORS AND TENSORS: Second Edition--Two Volumes Bound as One, Ray M. Bowen and C.-C. Wang. Convenient single-volume compilation of two texts offers both introduction and in-depth survey. Geared toward engineering and science students rather than mathematicians, it focuses on physics and engineering applications. 1976 edition. 560pp. 6 1/2 x 9 1/4. 0-486-46914-X

AN INTRODUCTION TO ORTHOGONAL POLYNOMIALS, Theodore S. Chihara. Concise introduction covers general elementary theory, including the representation theorem and distribution functions, continued fractions and chain sequences, the recurrence formula, special functions, and some specific systems. 1978 edition. 272pp. 5 3/8 x 8 1/2. 0-486-47929-3

ADVANCED MATHEMATICS FOR ENGINEERS AND SCIENTISTS, Paul DuChateau. This primary text and supplemental reference focuses on linear algebra, calculus, and ordinary differential equations. Additional topics include partial differential equations and approximation methods. Includes solved problems. 1992 edition. 400pp. 7 1/2 x 9 1/4. 0-486-47930-7

PARTIAL DIFFERENTIAL EQUATIONS FOR SCIENTISTS AND ENGINEERS, Stanley J. Farlow. Practical text shows how to formulate and solve partial differential equations. Coverage of diffusion-type problems, hyperbolic-type problems, elliptic-type problems, numerical and approximate methods. Solution guide available upon request. 1982 edition. 414pp. 6 1/8 x 9 1/4. 0-486-67620-X

A SURVEY OF INDUSTRIAL MATHEMATICS, Charles R. MacCluer. Students learn how to solve problems they'll encounter in their professional lives with this concise single-volume treatment. It employs MATLAB and other strategies to explore typical industrial problems. 2000 edition. 384pp. 5 3/8 x 8 1/2. 0-486-47702-9

NUMBER SYSTEMS AND THE FOUNDATIONS OF ANALYSIS, Elliott Mendelson. Geared toward undergraduate and beginning graduate students, this study explores natural numbers, integers, rational numbers, real numbers, and complex numbers. Numerous exercises and appendixes supplement the text. 1973 edition. 368pp. 5 3/8 x 8 1/2. 0-486-45792-3

A FIRST LOOK AT NUMERICAL FUNCTIONAL ANALYSIS, W. W. Sawyer. Text by renowned educator shows how problems in numerical analysis lead to concepts of functional analysis. Topics include Banach and Hilbert spaces, contraction mappings, convergence, differentiation and integration, and Euclidean space. 1978 edition. 208pp. 5 3/8 x 8 1/2. 0-486-47882-3

FRACTALS, CHAOS, POWER LAWS: Minutes from an Infinite Paradise, Manfred Schroeder. A fascinating exploration of the connections between chaos theory, physics, biology, and mathematics, this book abounds in award-winning computer graphics, optical illusions, and games that clarify memorable insights into self-similarity. 1992 edition. 448pp. 6 1/8 x 9 1/4. 0-486-47204-3

SET THEORY AND THE CONTINUUM PROBLEM, Raymond M. Smullyan and Melvin Fitting. A lucid, elegant, and complete survey of set theory, this three-part treatment explores axiomatic set theory, the consistency of the continuum hypothesis, and forcing and independence results. 1996 edition. 336pp. 6 x 9. 0-486-47484-4

DYNAMICAL SYSTEMS, Shlomo Sternberg. A pioneer in the field of dynamical systems discusses one-dimensional dynamics, differential equations, random walks, iterated function systems, symbolic dynamics, and Markov chains. Supplementary materials include PowerPoint slides and MATLAB exercises. 2010 edition. 272pp. 6 1/8 x 9 1/4. 0-486-47705-3

ORDINARY DIFFERENTIAL EQUATIONS, Morris Tenenbaum and Harry Pollard. Skillfully organized introductory text examines origin of differential equations, then defines basic terms and outlines general solution of a differential equation. Explores integrating factors; dilution and accretion problems; Laplace Transforms; Newton's Interpolation Formulas, more. 818pp. 5 3/8 x 8 1/2. 0-486-64940-7

MATROID THEORY, D. J. A. Welsh. Text by a noted expert describes standard examples and investigation results, using elementary proofs to develop basic matroid properties before advancing to a more sophisticated treatment. Includes numerous exercises. 1976 edition. 448pp. 5 3/8 x 8 1/2. 0-486-47439-9

Mathematics–Algebra and Calculus

VECTOR CALCULUS, Peter Baxandall and Hans Liebeck. This introductory text offers a rigorous, comprehensive treatment. Classical theorems of vector calculus are amply illustrated with figures, worked examples, physical applications, and exercises with hints and answers. 1986 edition. 560pp. 5 3/8 x 8 1/2. 0-486-46620-5

ADVANCED CALCULUS: An Introduction to Classical Analysis, Louis Brand. A course in analysis that focuses on the functions of a real variable, this text introduces the basic concepts in their simplest setting and illustrates its teachings with numerous examples, theorems, and proofs. 1955 edition. 592pp. 5 3/8 x 8 1/2. 0-486-44548-8

ADVANCED CALCULUS, Avner Friedman. Intended for students who have already completed a one-year course in elementary calculus, this two-part treatment advances from functions of one variable to those of several variables. Solutions. 1971 edition. 432pp. 5 3/8 x 8 1/2. 0-486-45795-8

METHODS OF MATHEMATICS APPLIED TO CALCULUS, PROBABILITY, AND STATISTICS, Richard W. Hamming. This 4-part treatment begins with algebra and analytic geometry and proceeds to an exploration of the calculus of algebraic functions and transcendental functions and applications. 1985 edition. Includes 310 figures and 18 tables. 880pp. 6 1/2 x 9 1/4. 0-486-43945-3

BASIC ALGEBRA I: Second Edition, Nathan Jacobson. A classic text and standard reference for a generation, this volume covers all undergraduate algebra topics, including groups, rings, modules, Galois theory, polynomials, linear algebra, and associative algebra. 1985 edition. 528pp. 6 1/8 x 9 1/4. 0-486-47189-6

BASIC ALGEBRA II: Second Edition, Nathan Jacobson. This classic text and standard reference comprises all subjects of a first-year graduate-level course, including in-depth coverage of groups and polynomials and extensive use of categories and functors. 1989 edition. 704pp. 6 1/8 x 9 1/4. 0-486-47187-X

CALCULUS: An Intuitive and Physical Approach (Second Edition), Morris Kline. Application-oriented introduction relates the subject as closely as possible to science with explorations of the derivative; differentiation and integration of the powers of x; theorems on differentiation, antidifferentiation; the chain rule; trigonometric functions; more. Examples. 1967 edition. 960pp. 6 1/2 x 9 1/4. 0-486-40453-6

ABSTRACT ALGEBRA AND SOLUTION BY RADICALS, John E. Maxfield and Margaret W. Maxfield. Accessible advanced undergraduate-level text starts with groups, rings, fields, and polynomials and advances to Galois theory, radicals and roots of unity, and solution by radicals. Numerous examples, illustrations, exercises, appendixes. 1971 edition. 224pp. 6 1/8 x 9 1/4. 0-486-47723-1

Mathematics-Probability and Statistics

BASIC PROBABILITY THEORY, Robert B. Ash. This text emphasizes the probabilistic way of thinking, rather than measure-theoretic concepts. Geared toward advanced undergraduates and graduate students, it features solutions to some of the problems. 1970 edition. 352pp. 5 3/8 x 8 1/2. 0-486-46628-0

PRINCIPLES OF STATISTICS, M. G. Bulmer. Concise description of classical statistics, from basic dice probabilities to modern regression analysis. Equal stress on theory and applications. Moderate difficulty; only basic calculus required. Includes problems with answers. 252pp. 5 5/8 x 8 1/4. 0-486-63760-3

OUTLINE OF BASIC STATISTICS: Dictionary and Formulas, John E. Freund and Frank J. Williams. Handy guide includes a 70-page outline of essential statistical formulas covering grouped and ungrouped data, finite populations, probability, and more, plus over 1,000 clear, concise definitions of statistical terms. 1966 edition. 208pp. 5 3/8 x 8 1/2. 0-486-47769-X

GOOD THINKING: The Foundations of Probability and Its Applications, Irving J. Good. This in-depth treatment of probability theory by a famous British statistician explores Keynesian principles and surveys such topics as Bayesian rationality, corroboration, hypothesis testing, and mathematical tools for induction and simplicity. 1983 edition. 352pp. 5 3/8 x 8 1/2. 0-486-47438-0

INTRODUCTION TO PROBABILITY THEORY WITH CONTEMPORARY APPLICATIONS, Lester L. Helms. Extensive discussions and clear examples, written in plain language, expose students to the rules and methods of probability. Exercises foster problem-solving skills, and all problems feature step-by-step solutions. 1997 edition. 368pp. 6 1/2 x 9 1/4. 0-486-47418-6

CHANCE, LUCK, AND STATISTICS, Horace C. Levinson. In simple, nontechnical language, this volume explores the fundamentals governing chance and applies them to sports, government, and business. "Clear and lively ... remarkably accurate." – *Scientific Monthly*. 384pp. 5 3/8 x 8 1/2. 0-486-41997-5

FIFTY CHALLENGING PROBLEMS IN PROBABILITY WITH SOLUTIONS, Frederick Mosteller. Remarkable puzzlers, graded in difficulty, illustrate elementary and advanced aspects of probability. These problems were selected for originality, general interest, or because they demonstrate valuable techniques. Also includes detailed solutions. 88pp. 5 3/8 x 8 1/2. 0-486-65355-2

EXPERIMENTAL STATISTICS, Mary Gibbons Natrella. A handbook for those seeking engineering information and quantitative data for designing, developing, constructing, and testing equipment. Covers the planning of experiments, the analyzing of extreme-value data; and more. 1966 edition. Index. Includes 52 figures and 76 tables. 560pp. 8 3/8 x 11. 0-486-43937-2

Mathematics-Geometry and Topology

PROBLEMS AND SOLUTIONS IN EUCLIDEAN GEOMETRY, M. N. Aref and William Wernick. Based on classical principles, this book is intended for a second course in Euclidean geometry and can be used as a refresher. More than 200 problems include hints and solutions. 1968 edition. 272pp. 5 3/8 x 8 1/2. 0-486-47720-7

TOPOLOGY OF 3-MANIFOLDS AND RELATED TOPICS, Edited by M. K. Fort, Jr. With a New Introduction by Daniel Silver. Summaries and full reports from a 1961 conference discuss decompositions and subsets of 3-space; n-manifolds; knot theory; the Poincaré conjecture; and periodic maps and isotopies. Familiarity with algebraic topology required. 1962 edition. 272pp. 6 1/8 x 9 1/4. 0-486-47753-3

POINT SET TOPOLOGY, Steven A. Gaal. Suitable for a complete course in topology, this text also functions as a self-contained treatment for independent study. Additional enrichment materials make it equally valuable as a reference. 1964 edition. 336pp. 5 3/8 x 8 1/2. 0-486-47222-1

INVITATION TO GEOMETRY, Z. A. Melzak. Intended for students of many different backgrounds with only a modest knowledge of mathematics, this text features self-contained chapters that can be adapted to several types of geometry courses. 1983 edition. 240pp. 5 3/8 x 8 1/2. 0-486-46626-4

TOPOLOGY AND GEOMETRY FOR PHYSICISTS, Charles Nash and Siddhartha Sen. Written by physicists for physics students, this text assumes no detailed background in topology or geometry. Topics include differential forms, homotopy, homology, cohomology, fiber bundles, connection and covariant derivatives, and Morse theory. 1983 edition. 320pp. 5 3/8 x 8 1/2. 0-486-47852-1

BEYOND GEOMETRY: Classic Papers from Riemann to Einstein, Edited with an Introduction and Notes by Peter Pesic. This is the only English-language collection of these 8 accessible essays. They trace seminal ideas about the foundations of geometry that led to Einstein's general theory of relativity. 224pp. 6 1/8 x 9 1/4. 0-486-45350-2

GEOMETRY FROM EUCLID TO KNOTS, Saul Stahl. This text provides a historical perspective on plane geometry and covers non-neutral Euclidean geometry, circles and regular polygons, projective geometry, symmetries, inversions, informal topology, and more. Includes 1,000 practice problems. Solutions available. 2003 edition. 480pp. 6 1/8 x 9 1/4. 0-486-47459-3

TOPOLOGICAL VECTOR SPACES, DISTRIBUTIONS AND KERNELS, François Trèves. Extending beyond the boundaries of Hilbert and Banach space theory, this text focuses on key aspects of functional analysis, particularly in regard to solving partial differential equations. 1967 edition. 592pp. 5 3/8 x 8 1/2. 0-486-45352-9

Mathematics–History

THE WORKS OF ARCHIMEDES, Archimedes. Translated by Sir Thomas Heath. Complete works of ancient geometer feature such topics as the famous problems of the ratio of the areas of a cylinder and an inscribed sphere; the properties of conoids, spheroids, and spirals; more. 326pp. 5 3/8 x 8 1/2.
0-486-42084-1

THE HISTORICAL ROOTS OF ELEMENTARY MATHEMATICS, Lucas N. H. Bunt, Phillip S. Jones, and Jack D. Bedient. Exciting, hands-on approach to understanding fundamental underpinnings of modern arithmetic, algebra, geometry and number systems examines their origins in early Egyptian, Babylonian, and Greek sources. 336pp. 5 3/8 x 8 1/2.
0-486-25563-8

THE THIRTEEN BOOKS OF EUCLID'S ELEMENTS, Euclid. Contains complete English text of all 13 books of the Elements plus critical apparatus analyzing each definition, postulate, and proposition in great detail. Covers textual and linguistic matters; mathematical analyses of Euclid's ideas; classical, medieval, Renaissance and modern commentators; refutations, supports, extrapolations, reinterpretations and historical notes. 995 figures. Total of 1,425pp. All books 5 3/8 x 8 1/2.
Vol. I: 443pp. 0-486-60088-2
Vol. II: 464pp. 0-486-60089-0
Vol. III: 546pp. 0-486-60090-4

A HISTORY OF GREEK MATHEMATICS, Sir Thomas Heath. This authoritative two-volume set that covers the essentials of mathematics and features every landmark innovation and every important figure, including Euclid, Apollonius, and others. 5 3/8 x 8 1/2.
Vol. I: 461pp. 0-486-24073-8
Vol. II: 597pp. 0-486-24074-6

A MANUAL OF GREEK MATHEMATICS, Sir Thomas L. Heath. This concise but thorough history encompasses the enduring contributions of the ancient Greek mathematicians whose works form the basis of most modern mathematics. Discusses Pythagorean arithmetic, Plato, Euclid, more. 1931 edition. 576pp. 5 3/8 x 8 1/2. 0-486-43231-9

CHINESE MATHEMATICS IN THE THIRTEENTH CENTURY, Ulrich Libbrecht. An exploration of the 13th-century mathematician Ch'in, this fascinating book combines what is known of the mathematician's life with a history of his only extant work, the Shu-shu chiu-chang. 1973 edition. 592pp. 5 3/8 x 8 1/2. 0-486-44619-0

PHILOSOPHY OF MATHEMATICS AND DEDUCTIVE STRUCTURE IN EUCLID'S ELEMENTS, Ian Mueller. This text provides an understanding of the classical Greek conception of mathematics as expressed in Euclid's Elements. It focuses on philosophical, foundational, and logical questions and features helpful appendixes. 400pp. 6 1/2 x 9 1/4. 0-486-45300-6